RIVERS OF DIAMONDS:

An Alluvial History of the Lower Vaal Basin, South Africa

by

David M. Helgren
University of California at Davis

THE UNIVERSITY OF CHICAGO
DEPARTMENT OF GEOGRAPHY
RESEARCH PAPER NO. 185

1979

Published 1979 by The Department of Geography
The University of Chicago, Chicago, Illinois

Library of Congress Cataloging in Publication Data

Helgren, David M 1947–

Rivers of diamonds.

(Research paper–University of Chicago, Department of Geography; no. 185)

Bibliography: p. 366

1. Geomorphology–South Africa–Vaal River watershed. 2. Sediments (Geology)–South Africa–Vaal River watershed. I. Title. II. Series: Chicago. Dept. of Geography. Research paper; no. 185.

H31.C514 no. 185 [GB440.S6] 910s [551.4′83′0968]

ISBN 0-89065-092-6

79-17790

Research Papers are available from:

The University of Chicago
Department of Geography
5828 S. University Avenue
Chicago, Illinois 60637
Price: $8.00 list; $6.00 series subscription

For Adrii, Katriena and Karin

PREFACE

This monograph describes the geomorphic history of the rivers in the Lower Vaal Basin, South Africa. The streams in this region, like those in many other semiarid, subtropical landscapes, flow in valleys that contain terraces and late Cenozoic alluvial deposits. The historical record in the Lower Vaal Basin is particularly impressive, since ancient alluvia are both abundant and well-exposed. In addition, many of the modern streams follow exhumed valleys of Paleozoic age.

River terraces are among the oldest of topics in geomorphology. Yet the last few decades of productive research in fluvial geomorphology have achieved little progress on the host of relevant questions. Recent research seems to have focused on only a few categories of streams and then too often from the ahistorical, engineering perspective. By building on empirical regularities to explain modern stream behavior in engineering terms, developmental problems have been neglected. A similar opportunity to compare modern streams with their long-term depositional residues and erosional forms seems to have been missed by historical geology. In particular, the late Cenozoic in general and the analysis of unconformities--the buried, relict slopes which record most of earth history--in particular deserve more detailed study.

In preparing this report, four issues have emerged as crucial in understanding the histories of rivers. Firstly, it is necessary to understand the volumes and textures of the sediments supplied to the stream valley. Secondly, the periodicities of sediment delivery, both into the valley and down the valley must be understood.

Thirdly, the character of fluvial discharge into and down the valley need to be determined. And fourthly, the periodicity of formative discharges must be assessed. An historical analysis, incorporating these issues in preliminary fashion, is included in the concluding chapter of this study.

Many people and institutions have assisted in the completion of this report. Their help is gratefully acknowledged. The project had its origin during a field season with Karl W. Butzer in 1970, when Dr. and Mrs. G. Fock kindly introduced me to the Lower Vaal Basin in the course of a week-long visit. During 1972-73 and 1975 the McGregor Museum (Kimberley) provided general logistical support and accommodations, and the museum director, Dr. R. Liversidge, and staff were particularly helpful. Special appreciation is due the many farmers and diggers who cheerfully granted me access to their properties and claims. Muriel Macey of the Kimberley Library provided references on the early travelers to the region, and J. B. Hawthorne and the Geology Department at De Beers Consolidated Mines kindly shared useful, unpublished data.

Most of the sediment samples were analyzed at the University of Chicago, where Dan Bowman and Chuck Kaplan frequently helped to take hydrometer readings. Additional samples were analyzed with the assistance of Yuri Werner at the University of Toronto. Alan Jopling of the University of Toronto gave excellent advice on several key issues of alluvial sediment transport. Sheila Boutcher (Toronto) described many of the thin sections. Most of the long calculations were completed at the University of Toronto Computer Centre.

My graduate studies at the University of Chicago were supported by a National Science Foundation Traineeship, while the 1972-73 field studies were supported by a Ford Foundation Stipend in Archeology. After completion of my doctoral program, an additional field season in 1975 was supported by a faculty research grant from the University of California, Davis, and by the L.S.B. Leakey Foundation. Liz Dutzi,

Matt Gillespie, Adrii Helgren, Bill Hollingsworth and Pam Ricci undertook the quality drafting and cartography that appear on the pages below. My thanks also go to Judy Riddle for typing the manuscript.

Special acknowledgement is due Karl W. Butzer, whose encouragement and guidance proved invaluable in all phases of this research.

David M. Helgren
Davis, California
December, 1978

TABLE OF CONTENTS

LIST OF TABLES

LIST OF FIGURES

Figure		Page

Figure Page

Figure | Page

CHAPTER 1

INTRODUCTION

> But there the veld is. And it is not good to think too much about it. For then it can lead you in strange ways.
>
> --Herman Charles Bosman
> in Veld Maiden

The Vaal Basin

From its headwaters above the 1700 m contour on the high plains of the southeastern Transvaal the Vaal flows westerly across the undulating grasslands of the High Veld before turning southerly and passing into the mesaform, steppe country along its lower course (see Fig. 1-1). Below Douglas in northern Cape Province the Vaal completes its 1350 km journey at an elevation of 970 m as it empties into the great river of South Africa, the Orange.

The continuing erosion in the Vaal basin was well underway before the fractionation of Gondwanaland in the later Mesozoic. The Vaal and its tributaries have spent the Cenozoic clearing away the near-horizontal sedimentary rocks of the Karoo System (Late Paleozoic-early Mesozoic) to create plains that arc from horizon to horizon. In addition, terrains in Precambrian lithologies beneath the Karoo rocks--the Pre-Karoo Surface--have been exhumed. The most famous product of this erosion is the High

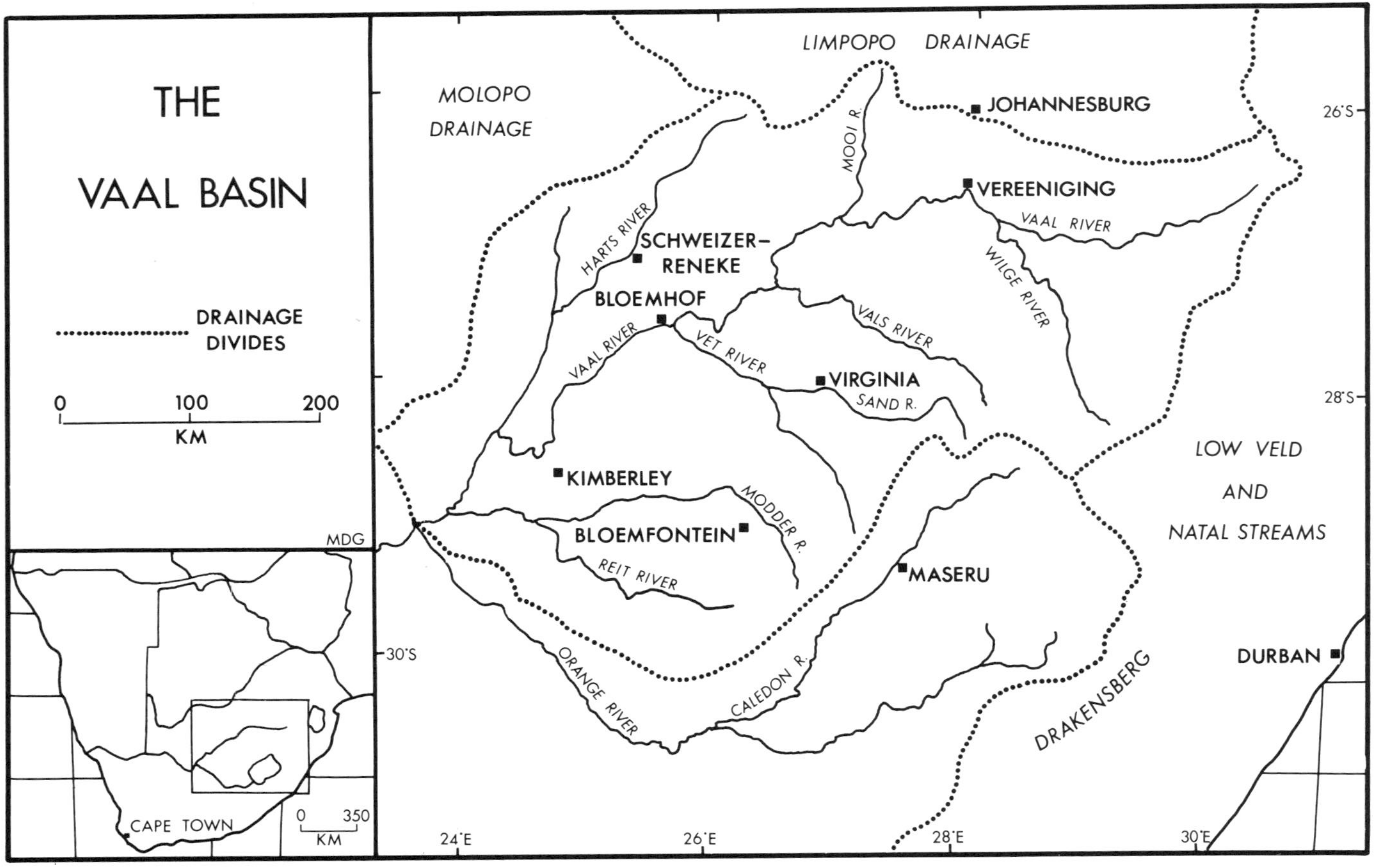

Fig. 1-1. The Vaal Basin.

Veld, which now dominates the landscapes of the southern Transvaal and Orange Free State. Starting near the 1200 m contour in the west and reaching above 1800 m at the base of the Lesotho highlands, the High Veld is composed of coalescent planar surfaces resting on Karoo sandstones and shales as well as Precambrian volcanics and metasedimentaries.[1] In addition parts of the High Veld are early Paleozoic plains of the Pre-Karoo Surface (Wellington, 1937; King, 1963: 228). Relief on the High Veld is provided by isolated low ridges and buttes (koppies) where hills of the Pre-Karoo Surface protrude through the Karoo covermass, or where resistant lithologies (terminal Karoo dolerites or sandstones) serve as caprocks. Below the High Veld is the Lower Vaal Basin where large areas of the Pre-Karoo Surface have been exposed and greater local relief is provided by more closely spaced ridges and koppies. For the purpose of this study the Vaal basin is formally divided into these upper and lower segments--the High Veld and the Lower Vaal Basin, respectively.

In this context of the planar landscapes, the boundaries of the Vaal basin expectably are poorly demarcated. To the north along the Limpopo watershed are the hills and upland plains that mark the Witwatersrand. To the east the watershed with a collection

---

[1]The term "High Veld" (Afrikaans: Hoë Veld) is a cultural toponymn dating back to the settlers of the nineteenth century, and specifically refers to the physical setting of undulating plains which extend across most of the Orange Free State and southern Transvaal. W. M. Davis (1906) first gave the High Veld a geomorphic definition, describing its gentle convex surfaces and isolated ridges as a monadnock-studded peneplain. Later the High Veld became a cornerstone of the landscape development theories of L. C. King (1963, 1967). The physiography of South Africa is described at length by Wellington (1955) and King (1963). The geology is summarized by du Toit (1954) and Haughton (1969).

of Natal and Transvaal lowveld streams is along the broad plateaus above the crest of the Drakensberg Escarpment, while the Orange-Vaal drainage divide is equally gentle, being marked by sequence of low-angle, erosional ramps in the southern Orange Free State. But the divide to the north and west with the Molopo drainage is the most subtle of all, where it curves indistinctly across low thresholds in karstic terrains. During the Tertiary the Vaal basin lost area to all these neighboring drainages as headwater captures reduced Vaal discharges (see Chapters 4 and 10 below). The Vaal once tapped the crest of the Drakensberg, and perhaps reached into the northwestern Transvaal and eastern Botswana (Williams, 1932). Although of great economic and cultural significance, drainageways remain unimpressive throughout the Vaal basin. The Vaal channel is seldom more than 100 m wide, and the adjacent floodways are proportionally narrow (Figs. 1-2 and 1-3). Even many of the main tributaries are less than 40 m wide.

Proceeding down the Vaal from its headwater region, the modern climate of the Vaal drainage becomes progressively more arid and warmer (see Schulze, 1965; and Preston-Whyte, 1974, for summaries). Downbasin the rainfall reliability decreases, while the intensity of the infrequent rains increases. Rainfall throughout the Vaal basin is concentrated in the summer months (January, February, March). Summer rainfalls most commonly occur when a large, low-pressure cell becomes centered over the sub-continent, and moisture is drawn over the Drakensberg into the basin from the Indian Ocean to the east. The extent, duration, and regularity of these largely convectional rains decrease inland. High-pressure cells dominate the Vaal drainage in the winter and rare rains occur only when these sluggish highs weaken. Annual average precipitation in the Vaal drainage decreases from 800 mm upstream to 300 mm downstream. Although the Vaal basin is centered on

Fig. 1-2. The Vaal River near its source on the plains of the southeastern Transvaal.

Fig. 1-3. The Vaal River at Vereeninging, nearly 300 km from its source.

latitude 28° south, the high elevation maintains relatively cool temperatures. The occurrence of light frost on the High Veld averages 120 to 150 days per year, but snows are rare and no frozen soils occur. In the Lower Vaal Basin the incidence of light frost averages 100 days per year. January (summer) mean temperatures on the High Veld range from 18° to 23° C, while July (winter) mean temperatures range from 7° to 9° C. In the lower basin temperatures are slightly higher (see Chap. 2).

The modern vegetation of the Vaal basin reflects edaphic conditions as much as climate. On the High Veld, weathering of Karoo sedimentaries produces well-developed, medium to coarse-textured soils, which once supported impressive grasslands. Arboreal vegetation is uncommon, but is found along water courses, around springs, and on moister soils in some topographic depressions. In the Lower Vaal Basin biogeographic history becomes more important and the vegetation patterning demonstrates the transitional position of this area between the High Veld grasslands to the east, the Kalahari thorn-savannas to the north and west, and the semidesert shrub of the Karoo to the south (see Acocks, 1953; van Rooyen and Burger, 1974a; Chap. 2 below).

Traditionally the Vaal system seems to have responded to this environmental mosaic with erratic summer floods and only minor base flows. However, since the 1920's several large dams and many small weirs have been constructed, and discharge in the modern Vaal system has come to be more and more regulated. Now the Vaal network consists of a series of ponded reaches, and the former, highly seasonal flows have been sharply reduced and redistributed throughout the year. Before modification, the Vaal sometimes ceased to flow along much of its lower course during the dry season, but now a perennial supply of water is assured for agricultural and industrial users throughout the basin.

Reconstructing the fluvial processes effective in channel maintenance and sediment transport prior to dam and weir construction is now a serious, if not unsolvable problem in both the historiography of the few old flow records and in the geology of the most recent alluvial deposits. Above Vereeniging, and after its confluence with the important Orange Free tributary, the Wilge, the Vaal appears to have been virtually perennial, but the possibility of the main channel seasonally stagnating increased steadily downstream from there. Between Vereeniging and Warrenton the Vaal receives several important tributaries, particularly the Vals, Vet-Sand, and Mooi rivers. The Vals, Vet, and other leftbank tributaries rise on the plains of the High Veld; prior to modification their discharge was largely dependent on passing rainstorms, and fluvial activity could be severely reduced not only during the dry season, but between individual summer rainy periods. On the other hand, the Mooi and several other northbank tributaries rise in the karstic terrain of the western Witwatersrand and receive a perennial base flow from the dolomitic bedrock; rainfall runoff, and normal groundwater periodicities are superimposed on this flow curve (see Wellington, 1955: 266-273).

Between Warrenton and the Orange confluence the Vaal receives only two main tributaries, the Riet and Harts rivers. The Riet-Modder system drains much of the central and southern High Veld, and with modern flow management, maintains low, perennial flow. The Harts, like the Mooi, drains a karstic terrain, but precipitation yield is low and the unimpressive channel is generally less than 20 m wide. The lower reaches of the Vaal originally appear to have lacked significant through drainage except during the summer flood season. A minor flood in one High Veld reach may have only had a minor effect downstream, with major system-wide floods predicated on protracted rains after groundwater recharge through most of the basin. The image persists that prior to dam and weir construction, flow in any one reach of the Vaal network was largely dependent on rainfall and groundwater discharge in the local catchment, with rainfall in distant areas upbasin of potentially limited, local significance.

Unlike the Vaal, the Orange system drains highlands along the western flanks of the Drakensberg. This source area has perennial precipitation as well as a major spring snow melt, maintaining an impressive flow throughout the year. Historic, summer floods on the Orange have been devastating.

Historical Background of This Study

The spatial context for this report is the Lower Vaal Basin, defined here as the drainage of the Vaal below the Vaal-Harts wier near Warrenton. Included in the region are the Riet River drainage up to the confluence with the Modder River and the Harts drainage up to the Spitskop Dam.[2] Thus the northern boundary of the Lower Vaal Basin is at approximately 28°05' south latitude, while the eastern edge parallels the Orange Free State - Cape Province border. The western margin of the basin follows the Gaap Escarpment[3] down to the Orange. Also investigated was the sector of the Orange River from Hopetown, above the Vaal conflu-

---

[2] By implication the High Veld begins above Warrenton and the Modder-Riet confluence. This definition does not closely compare to King's mapping (1967: Fig. 119) which places the boundary at a contour 20 to 70 m higher.

[3] The 70 to 120 m high cliffs of the Gaap Escarpment (Afrikaans: Ghaap se Berg) mark the east-facing margin of the karstic Gaap Plateau (Afrikaans: Ghaap Plato). In the first report on the regional geology of the Lower Vaal Basin, Stow (1874) named the

ence, to the point where the arc of the Gaap escarpment, if extended, would cross the Orange. This is near Read's Drift, at about 29°12' south latitude and 23°23' east longitude (see Fig. 1-4).

The Lower Vaal Basin was settled by semi-nomadic pastoralists of European descent in the early nineteenth century. At this time Bushman bands still survived, and Korana, Hottentot, and Griqua herders sporadically roamed the region. In addition, Tswana clans maintained small but successful communities in the lower Harts valley. The early European travelers, most notably Campbell (1816, 1822), Burchell (1822-24), Thompson (1827), Steedman (1835), and Harris (1839), recount the daily rigors of their journeys and relate their fascination with the wildlife and the inhabitants. However, they tell us little about the physiography, and the exact routes of their journeys are often difficult to reconstruct. The travelers generally found little worth recounting on the semiarid plains, but they were singularly impressed by the green-fringed channel of the Orange and usually described each perilous crossing in detail. Yet their many crossings of the Vaal were only briefly noted. The Vaal was no hazard and apparently not worthy of commentary, compared to the Orange.

Gaap Escarpment the "Campbell Randt", a name championed by King but which never gained local acceptance. Campbell (1822, vol. 1: 50) called the south margin of the escarpment "White Hill", a toponymn which persists today in "Witberg", he also (Campbell, 1822, vol. 2: 42) named the Gaap Plateau the "Alers Plain". The most common English spelling in the last century for Gaap has been Kaap, which has proved hazardous since in Afrikaans, kaap = cape. Gaap is both a more ancient and a more popular term today. Van Vreeden (1961) indicates Gaap refers to a group of common, generally edible, small plants found on the plateau, particularly Trichocaulon barkly (the "Berg N'Gaap") and Stapelia jucunda (the "Ghapie"). He traces the lineage of the term as follows: Korana--≠ hab (≠ is the alvelor click) to early Griqua--≠ hāp to Afrikaans Griqua--Ghaap.

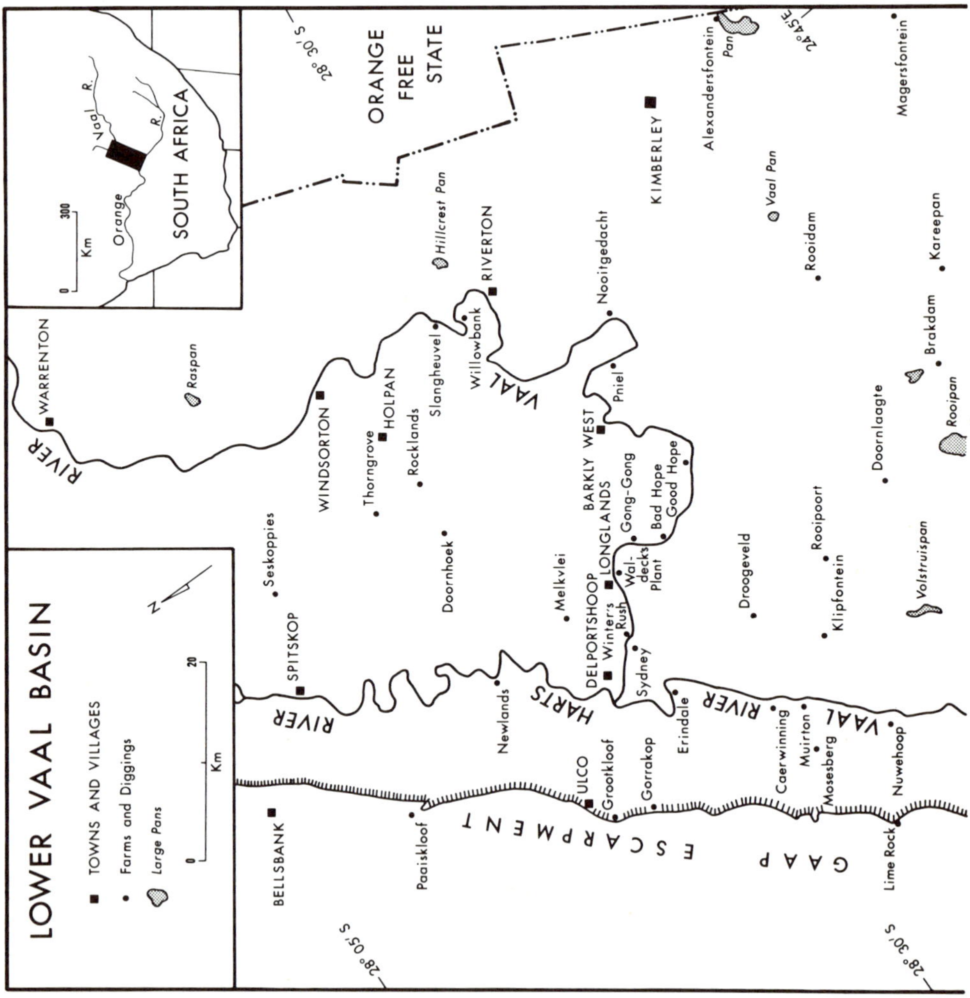
LOWER VAAL BASIN
TOWNS AND VILLAGES
Farms and Diggings
Large Pans
0
20
Km
N
0
300
Km
Vaal R.
Orange R.
SOUTH AFRICA
ORANGE FREE STATE
28° 30' S
28° 05' S
28° 30' S
24°45'E
WARRENTON
Raspan
VAAL RIVER
HARTS RIVER
WINDSORTON
HOLPAN
Thorngrove
Rocklands
Slangheuvel
Willowbank
Hillcrest Pan
RIVERTON
Nooitgedacht
KIMBERLEY
Alexandersfontein Pan
Magersfontein
Vaal Pan
Rooidam
Kareepan
Brakdam
Rooipan
Pniel
BARKLY WEST
Gong-Gong
Bad Hope
Good Hope
LONGLANDS
Wal-decks Plant
DELPORTSHOOP
Winters Rush
Sydney
Melkvlei
Doornhoek
Seskoppies
SPITSKOP
Newlands
Doornlaagte
Rooipoort
Droogeveld
Klipfontein
Volstruispan
Erindale
Caerwinning
Muirton
Mosesberg
Nuwehoop
Lime Rock
ULCO
Grootkloof
Gorrakop
Paaiskloof
BELLSBANK
GAAP ESCARPMENT

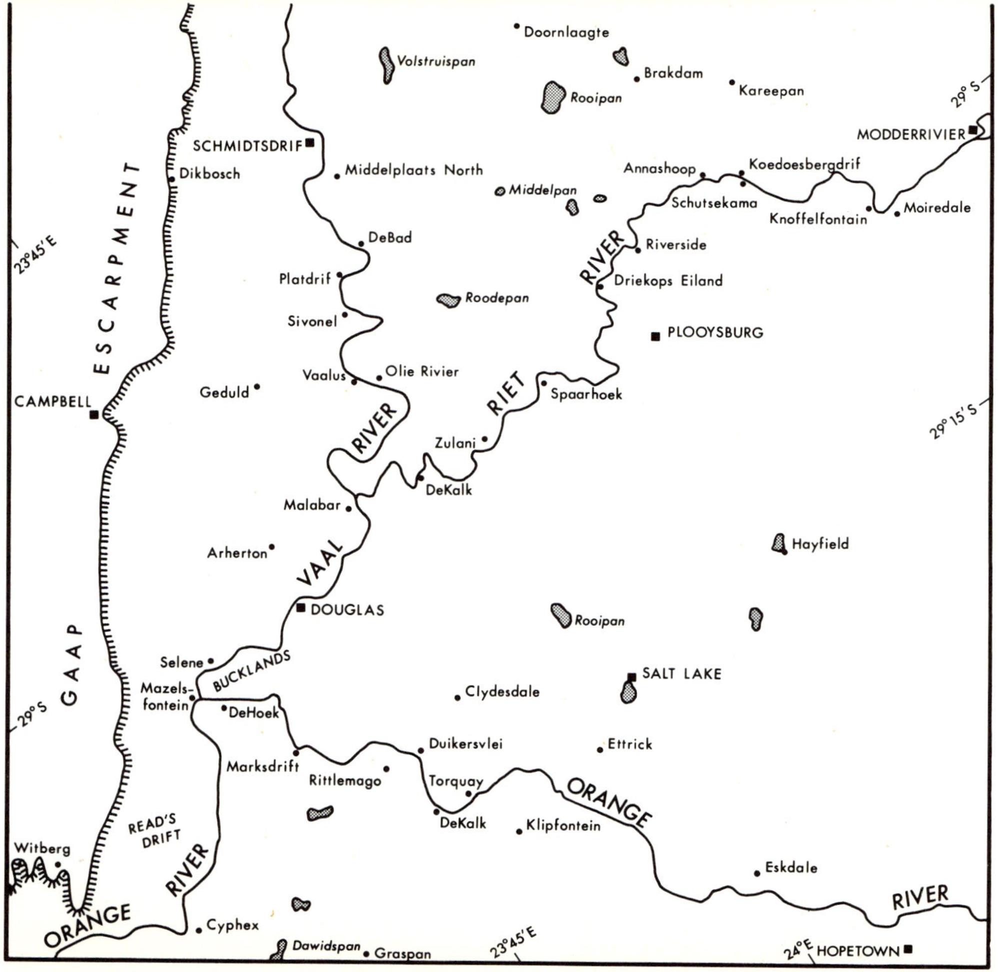

Fig. 1-4.
The Lower Vaal Basin

A few diamonds were found along the Orange near Hopetown in 1866, but the search for more diamonds was not immediately fruitful. Then in 1869 the 83.5 carat "Star of South Africa" was found[4] and the diamond rush was on. The surge of diamond hunters soon led to further discoveries and precipitated the rapid establishment of the South African diamond fields.[5] The rush of diggers was first directed to the alluvial gravel deposits, particularly those of the Vaal at Barkly West (Klipdrift, Pniel), Windsorton (Hebron), Gong-Gong, Longlands, and Sydney. However, digger interest in the "alluvials" waned in 1871 and 1872 after the discovery of the kimberlite pipes on the upland plains at Jagersfontein and Kimberley. The diamond fields of South Africa were a minor sensation in Victorian Europe, and the region was visited by many of the popular writers and travelers of the period (Trollope, 1878, is the best example). Like the earlier narrators these writers also directed their interest to the life in the boomtowns and the perils of travel for European gentlemen, while neglecting the physical landscape.[6]

---

[4]Local lore maintains that the "Star of South Africa" was found on the farm Sandfontein, located 25 km south of the Vaal-Orange confluence and at least 12 km from the nearest ancient alluvial deposit. This gem either was weathered from the local Dwyka tillite, or transported to the discovery site by Bushmen or an ostrich. Local lore could also be covering up a frontier fraud.

[5]On 27 October, 1871, the Lower Vaal Basin and areas to the west became a British territory under the name of Griqualand West.

[6]The best naturalist to visit the region during this period was Holub (1881). His account provides an interesting counterpart to those of Campbell and Burchell a half-century earlier.

The region has a long history of earth science research. The diamond discoveries sparked the first scientific inquiries into the alluvial deposits of the Lower Vaal Basin (see Wagner, 1914, for a bibliography). It was quickly recognized that the gravels represented former channel deposits of the Vaal River, but their significance for South African natural history was largely neglected. Studies were directed primarily to the economic problems of explaining the erratic occurrences of diamonds within the gravels and of finding the source-rocks for the alluvial diamonds. Highlights of earth science research during this period were the regional geological surveys of A. L. du Toit (1906, 1907, 1908, 1910). Besides diamonds, the diggers found nearly-as-rare archaeological and paleontological materials in the gravels. Brief reports of Stone Age implements were made by several of the scientific visitors and the debris they gathered was deposited in many of the world's museums. Even though the stone tools typically came from the secondary and tertiary contexts of diggers' waste heaps an artifact stratigraphy was eventually established (for a review see Goodwin, 1928, and van Riet Lowe, 1937). Upon finding fossil bones and teeth in their excavations, the diggers frequently saved them, and after a few years most of this skeletal material found its way to several museums. Again the original provenience of these mammalian remains is seldom known in any detail (Helgren, 1977). The first fossil described, the 'mastodon' of Beck (1906, now a gomphothere), continues to be one of the most enigmatic in the collection.

In the mid-1930's C. van Riet Lowe encouraged the first systematic investigation of the "Vaal River Gravels", as the ancient alluvia had come to be known. In one of the very first cooperative projects between archaeologists and earth scientists, van Riet Lowe, with P. G. Söhnge and D. J. L. Visser of the Union's Geological Survey, examined the Vaal River terraces and gravel deposits from near Vereeniging to Delportshoop. Their memoir (1937) was complemented by the publication of H. B. S. Cooke's 1949 report on the fossil mammals. While Söhnge and Visser were hesitant to reach any systematic geologic conclusions (see page 127 of the 1937 monograph), van Riet Lowe (summarized in 1952) went on to postulate a detailed alluvial-terrace chronology and a comprehensive sequence of climatic and cultural changes.

In briefest terms, van Riet Lowe identified three broad classes of alluvial deposits in the Lower Vaal Basin--(1) the Older Gravels, (2) the Younger Gravels, and (3) the succeeding "Calcified Sands", "Youngest Gravels", and "Recent Flood Silts". The Older Gravels generally are thin veneers of cobble-grade gravels which occur at descending elevations on the margins of the river valley. Van Riet Lowe thought they might contain pre-Acheulian artifacts. The Younger Gravels are thick accumulations of similar gravels, located on or near the modern valley bottom, and containing Acheulian artifacts and mammalian fossils. Van Riet Lowe believed there were three separate gravel aggradations in the Younger Gravels and that each was followed by an interval of "Calcified Sands" deposition. Still later, additional gravels were reworked in tributary gullies to become the "Youngest Gravels" while brown "Flood Silts" were deposited along the modern channel. Each depositional unit was supposed to contain a different category of stone tools. To explain his observations, van Riet Lowe correlated gravel deposition with a humid climate, and sand deposition with an arid climate, and then proceeded to match his apparent wet-dry-wet sequence and artifact stratigraphy with the then-popular, East African pluvial chronology. Further commentaries and refinements were offered by Cooke (1947) and Breuil, et al. (1948), and the sequence was introduced into the world literature by Zeuner (1945).

However, when this first impetus for research passed, the explanatory model championed by van Riet Lowe came to be regarded more and more equivocally. Modifications and even negative re-assessments were presented. Not only were objections raised concerning the terrace sequence and climatic interpretation (Flint, 1959; Maarleveld, 1960; Cooke, 1967; Bond, 1967; Partridge and Brink, 1967; Butzer, Helgren et al., 1973), but also of the archeological record (Cole, 1961; Mason, 1962, 1967; Helgren, 1978). The present study clarifies many of these issues.

The basis of this report was a detailed field investigation carried out from September, 1972, to May, 1973, and re-checked June to September, 1975. The field data were mapped to the base of 1:50,000 topographic maps, as corroborated by multi-scale, panchromatic aerial photos (at 1;30,000 to 1:60,000) and LANDSAT images. The terrace, hillslope, and stratification nomenclatures are deliberately simplified to eliminate, in as far as possible, genetic connotations and trivial or equivocal generic concepts. Terrace elevations were measured with an Abney level to modern low flow level (non-aquatic vegetation line) and typically are listed as ranges, reflecting upper and lower margins as well as the expectable irregularities introduced in initial deposition and subsequent erosion. Time-stratigraphic terms follow Haughton (1969) for the pre-Quaternary and Butzer and Isaac (1975) for the Quaternary. For the purpose of convenience, the archeological nomenclature follows van Riet Lowe (1952) with appropriate modifications after Sampson (1974).

Over 300 sediment and soil samples were collected and analyzed for textural, lithological, and chemical attributes. In addition, more than 30 samples were analyzed in thin section. The nomenclature, methods, and data are detailed in Appendices A and B.

Objectives of this Study

Altogether this report focuses on four interrelated issues. The primary theme is the stratigraphy and genesis of the alluvial fills and low terraces along the Lower Vaal River (Chaps. 5 to 8 and 10). These features, which include the Younger Gravels (the Rietputs Formation, see Chap. 5 below) and their capping fine-grained alluvia (the Riverton Formation, see Butzer, Helgren, et al., 1973) provide a chronology of Quaternary fluvial dynamism which is the most comprehensive known in the interior of southern Africa. The genesis of these features carries implications both for regional geomorphic history and for the context of the in situ archeological and paleontological remains (Helgren, 1977; 1978).

The second theme is provided by the stratigraphy and implications of the Older Gravels, ancient alluvia scattered across the uplands of the Lower Vaal Basin (Chap. 4). Similar deposits are present in most erosional landscapes, but those along the Vaal have been exposed by diamond diggers allowing a rare opportunity for detailed study. Together, the Older and Younger Gravels provide an unparalleled vision of the Cenozoic denudation in the Vaal basin, which must have implications for the development of both the High Veld and the modern landscapes of the Lower Vaal Basin.

The third theme is the lithological and structural setting for the Older and Younger Gravels (Chap. 3). This comprises a survey of regional tectonic history and detailed analysis of late-Paleozoic glaciation in the region. The Lower Vaal Basin contains the widest known areas of exhumed Paleozoic terrains. As elements of the modern landscape, this Pre-Karoo Surface deserves analysis in its own right. But its significance to the development of the later Cenozoic deposits has proved to be crucial.

The fourth theme (Chap. 9) centers on the matter of post-Karoo planation surfaces and the related landscape development theories of Lester C. King (1963, 1967). Using both simplified theoretical issues, and data from within and without the Lower Vaal Basin, it is apparent that King's paradigm is inadequate to explain either the gross geomorphologic configuration of the region, or the development of the Older and Younger Gravels in general.

None of these matters is exhausted here, but it is hoped that this report will serve to advance understanding of the environmental history of the Vaal basin across a broad front, providing guidelines for further research on the many related issues.

CHAPTER 2

SYNOPTIC PHYSICAL GEOGRAPHY OF THE LOWER VAAL BASIN

> We had now entered upon a new region,as the northern side of the Great River may justly be considered. The traveler may here remark a difference of character in the appearance and nature of the country,....
>
> --<u>W. J. Burchell</u> (1822, v. 1: 324) upon crossing the Orange below Read's Drift, 17 September, 1811

> The country we had now entered upon was very different from the dry and naked deserts we had lately traversed. The magnificent windings of the Craddock (Orange) were full of view; and to the west and north-west an extensive and almost boundless landscape opened out, adorned with thickets and enlivened with multitudes of large game.
>
> --<u>George Thompson</u> (1827: 64) on crossing the Orange near Hopetown, 6 June, 1823

Landscapes of the Lower Vaal Basin

One traditional goal of physical geography has been to describe and interpret a region prior to its modification by technically sophisticated societies. By filtering out the effects of cultural landscape modification, a conceptual model of a region can be constructed to serve as datum for the measurement of more ancient landscape changes. While research in physical geography during the last two decades has been directed toward detailed analysis and modeling of individual processes, regional synthesis remains necessary and relevant--especially if the region is to be viewed as a spatial array of historically interlocking relationships. Regional studies remain at the core of physical geography.

The semiarid landscapes of the Lower Vaal Basin have been modified to a limited extent since the mid-nineteenth century. The early travelers (e.g. Burchell, Campbell, Harris) report no "native towns"[1] in the region, but only ephemeral encampments which were moved as the immediate grazing and foraging resources were depleted. Since European settlement, the region has been fenced into farms, and a network of country villages established. Kimberley developed into the only true urban place while the much smaller villages are principally secondary distribution points for farmers and miners. The growth of the city and villages precipitated a relative "deforestation". Large numbers of _Acacia giraffae_ were cut in the

---

[1]Only two towns of note were present between the Orange and Kuruman Rivers at the turn of the nineteenth century. These were Klaarwater (Griquatown), which was settled by Coloured refugees from the Cape Colony, and Litakun (Lattakoo), a Tswana settlement near the present site of Kuruman. Burchell describes Klaarwater on 2 October, 1811, as "merely a row of half a dozen reed cottages" (1822, Vol. 1: 352), and Litakun on 13 July, 1812, as a ". . . Multitude of houses. . . . They occupied, in detached groups, a portion of the plain, not less than a mile and a half in diameter (1824, Vol. 2: 356).

savanna parklands to provide building materials and fuel for the Hopetown-Kimberley-Warrenton rail line, and mine-shaft supports (G. J. Fock, 1973: personal communication). Town and mine sites changed dramatically through construction and engineering works. Settlement resulted in the widespread substitution of the original large ungulates population with domesticated animals. Overgrazing has thinned the vegetation and favored xerophytic species (Acocks, 1953). Soil erosion has removed important parts of the fragile A-horizons through increased effectiveness of both eolian and fluvial processes. Clearing for irrigated agriculture on alluvial bottomlands and on the Vaal-Harts interfluve has largely destroyed the original vegetation. Despite these changes, much remains unchanged in the Lower Vaal Basin. The climate has fluctuated little, and the bedrock structures remain the same. The significance of man-caused floral and faunal changes on landforms appears minor, and, relative to this short temporal span, the shapes of the semiarid plains seem unalterable.

Climate

The climate of the Lower Vaal Basin is characterized by dry, cool winters, and warm to hot summers. Summer rainfall maxima are typically the product of sporadic thunderstorms. Like most semiarid settings, the amount of insolation is high throughout the year (e.g. Kimberley has only one or two "overcast" days per month during the summer; Schulze, 1965: 31). During the winter, weather patterns are usually dominated by a high pressure cell centered over the Drakensberg. Consequently no humid air masses can advance inland from the Indian Ocean. However this high weakens slightly with the passage of westerly low pressure cells along the south coast of Cape Province. During a particularly strong south coast low, the continental high may be briefly displaced and rains (or even snows) may be infrequently swept into the Lower Vaal Basin from the southwest. The other opportunity for rare winter precipitation in the basin occurs if the Drakensberg high becomes unusually shallow and sluggish. Then a south coast low of typical intensity will extend further inland than usual and may cause the winds of the Drakensberg anticyclone to rapidly accelerate, perhaps bringing a moist air mass from the Moçambique channel across the northern Transvaal. In this situation light rains may spread across Griqualand West -- a product of either the high pressure flow to the north or the low to the south. During the summer, weather patterns are usually dominated by an isolated low pressure cell situated over Botswana. Most of the precipitation in the Lower Vaal Basin occurs when this low deepens and extends southward. As a result moist air is brought into the basin from the Indian Ocean to the east, and thundershowers periodically spread over the region from northeast to southwest. More rarely the Botswana low may link with a south coast low (perhaps most commonly during the transitional seasons) and frontal precipitation advances into the region from the south. Since the Lower Vaal Basin is located near the center of the South African sub-continent, potential rain-bearing airmasses, whether from the east or south coasts, tend to be largely drained of moisture before they reach the region.

Climatic variation within the basin is not dramatic. During the winter, average temperatures decrease 1° to 2° C from north to south. Winter aridity is relatively uniform. During the summer, average temperatures increase 1° to 3° C and average rainfalls decrease 25 to 100 mm from northeast to southwest across the basin in response to decreasing elevations and decreasing thunderstorm activity (Table 2-1). The temperature statistics most relevant for rock weathering and biologic processes are readings at +5 cm and in the soil. These are recorded at Vaal-Harts, 20 km north of

TABLE 2-1

RAINFALL AND TEMPERATURE IN THE LOWER VAAL BASIN

		J	F	M	A	M	J	J	A	S	O	N	D	Year
Rainfall (monthly averages, mm)														
Warrenton	(48)[a]	56.6	65.1	78.9	36.6	16.6	8.4	6.8	9.9	10.8	28.8	37.1	56.2	411.8
Windsorton	(46)	59.7	58.2	72.1	37.8	18.7	5.9	6.0	10.9	10.0	26.8	38.9	52.4	397.4
Spitskop	(37)	55.5	59.5	71.6	29.4	13.1	5.4	3.3	7.2	9.0	20.9	29.0	49.0	352.9
Barkly West	(70)	58.8	68.2	76.8	40.9	16.9	7.7	6.1	9.9	10.4	25.2	39.4	56.9	417.2
Sydney	(43)	53.8	67.6	77.7	33.9	16.4	4.0	5.8	9.3	8.5	20.4	30.6	50.3	378.3
Schmidtsdrif	(35)	41.4	57.1	61.3	26.3	10.9	4.2	1.8	7.8	9.8	16.8	31.1	42.7	311.2
Modderrivier	(41)	37.2	42.6	62.7	32.2	12.1	2.5	5.4	8.0	11.5	25.6	31.0	33.0	303.8
Plooysburg	(28)	43.6	50.7	59.1	33.1	12.8	5.6	3.2	7.4	7.4	26.4	42.4	39.5	331.2
Douglas	(70)	46.3	60.0	69.3	29.1	16.7	4.3	3.7	6.8	10.4	18.5	26.9	33.7	325.7
Hopetown	(81)	40.7	46.1	62.0	31.9	20.3	6.8	5.9	7.5	11.6	19.4	31.2	29.6	313.0

Temperature (monthly average daily minima and maxima °C)

Kimberley	max	32.5	31.0	28.4	25.3	21.2	18.6	18.5	21.8	24.6	28.3	29.9	31.7	26.0
(1894-1950)[b]	min	17.8	17.0	14.8	10.9	6.4	2.8	2.4	5.1	8.2	12.1	14.2	16.4	10.7
Barkly West	max	32.9	31.2	29.2	25.8	22.4	19.2	19.0	22.2	24.7	29.2	30.4	32.7	26.6
(1926-1940)	min	16.9	16.2	14.0	9.7	5.0	1.3	0.4	3.3	7.0	12.1	14.0	16.2	9.7
Sydney	max	34.9	33.2	30.5	27.3	23.6	20.4	20.1	23.4	26.9	30.9	32.2	34.3	28.2
(1914-1950)	min	18.5	17.6	15.5	11.1	6.1	2.1	1.1	3.8	7.8	12.5	14.9	17.4	10.7
Hopetown	max	32.9	32.1	30.3	25.0	20.8	18.3	17.2	21.2	23.9	27.7	29.9	33.1	26.0
(1893-1916; 1937-1948)	min	18.3	17.2	14.6	10.3	5.9	2.3	0.6	3.5	6.9	10.7	13.7	16.3	10.0

[a]The numbers in parentheses refer to the number of years of record prior to 1961. The data are from Climate of South Africa, pt. 9 (1965).

[b]The years in parentheses refer to the period of observation. The data are from Climate of South Africa, pt. 1 (1954).

Warrenton, and at Rietrivier, 8 km south of the Riet-Modder confluence (Table 2-2). Variances between these stations probably reflect local air drainage patterns more than macro-climatic differences. While frost may be absent on north-facing slopes, winter frosts on flat areas and in topographic depressions may be frequent. However such frosts are relatively brief and light, due to the length and intensity of insolation even during the winter months, so that frost penetration is very shallow. The effect of frost on bare rock outcrops in the modern landscape is more difficult to assess. No angular breccias indicative of contemporary frost weathering were found in the region.

Rainfall at Kimberley averages 400 mm per year, nearly half falling during three summer months (Table 2-3). Noteworthy is the high variability in monthly rainfall from year to year, particularly during the nine driest months (April to December). However potential evaporation in Kimberley greatly outstrips precipitation, with over 2500 mm annually evaporating from a "class A" pan (Table 2-4). Furthermore, the greatest monthly average evaporation is in November and December -- prior to the peak period in rainfall activity.[2]

Surface winds at Kimberley show little seasonal variation. On the average the most common and strongest winds come from NW to NNE. In January north is both the most frequent and strongest wind direction, while in July north remains the most common direction, but the strongest winds come from NNW. The summer wind rose shows secondary peaks from southwest to west; these vectors may be the product of the passage of convective disturbances and of individual thunderstorms. The highest average winds occur during late winter and spring. Typical diurnal variations in windspeed and direction

---

[2]These evaporation maxima coincide with the period of maximum "windiness".

TABLE 2-2

GRASS HEIGHT AND SOIL TEMPERATURES (C°) AT RIETRIVIER AND VAALHARTS

		J	F	M	A	M	J	J	A	S	O	N	D	Year
Rietrivier (29°07' S, 24°36' E)														
+5 cm	$\bar{x}$	13.9	13.9	12.0	7.6	2.4	-0.7	-0.7	0.1	4.0	7.1	11.3	13.7	7.1
(Grass)	max	20.2	20.2	19.4	16.0	12.3	8.4	8.6	10.7	13.2	16.2	20.4	19.7	15.4
	min	0.2	5.6	3.1	-1.0	-5.5	-9.8	-9.3	-7.6	-5.6	-4.2	4.1	7.8	-1.8
-10 cm (Bare) 0800 hrs	$\bar{x}$	24.4	23.3	21.6	15.9	10.2	6.8	7.2	9.2	13.6	17.7	21.1	23.6	16.2
-10 cm (Bare) 1400 hrs	$\bar{x}$	34.0	33.9	30.8	24.3	18.2	14.5	14.9	18.0	23.4	28.1	32.5	33.8	25.5
Vaalharts (27°57' S, 24°50' E)														
+5 cm	$\bar{x}$	13.3	14.1	11.9	6.4	0.7	-2.8	-1.8	-1.0	2.5	6.8	11.0	12.7	6.1
(Grass)	max	19.3	19.4	18.5	15.3	13.0	6.2	9.1	13.6	16.7	16.6	19.6	19.1	15.5
	min	5.0	5.2	0.5	-3.0	-6.6	-12.0	-10.6	-11.0	-8.0	-6.1	-0.3	4.0	3.5
-10 cm (Bare) 0800 hrs	$\bar{x}$	23.8	23.2	21.3	16.7	11.2	7.6	7.9	9.9	14.1	18.3	21.8	23.0	16.6
-10 cm (Bare) 1400 hrs	$\bar{x}$	35.0	33.9	31.4	25.9	19.4	15.1	15.5	18.8	24.6	29.5	33.6	34.2	26.4

[a]These data are for a period of five years. The grass height max and min are the respective maximum and minimum for these months during this interval. Data are from Schulze (1965, Tables 31 and 38).

TABLE 2-3

RAINFALL VARIABILITY AT KIMBERLEY, 1877-1969[a]

	J	F	M	A	M	J	J	A	S	O	N	D	Year
Rainfall (mm)													
$\bar{x}$[b]	59.6	64.4	73.9	37.8	19.6	7.7[c]	6.7	8.4[d]	13.5	27.8	42.8	52.1	412.2
σ	43.7	42.7	41.8	31.6	19.6	14.6	11.2	16.1	26.9	24.8	35.3	41.1	127.7
CV%	73.3	66.2	56.6	83.8	99.9	188.8	166.3	191.6	198.9	89.2	82.5	78.8	31.0
Days With Rain													
$\bar{x}$	8.2	8.4	9.4	6.0	3.7	1.7	1.4	1.7	2.0	4.5	6.0	7.4	60.7
σ	3.7	3.4	3.9	3.2	2.7	1.8	1.9	1.8	2.4	2.8	3.0	3.9	16.9
CV%	45.1	41.1	41.5	52.5	72.7	106.8	133.8	105.4	117.0	62.0	49.9	52.5	27.9

[a]Data source is unpublished mimeograph from the Kimberley weather office. Rainfall for the years 1877 to 1896 was observed at Kimberley, for the years 1897 to 1929 and 1949 to 1969 at Kenilworth, and for the years 1929 to 1948 at the Meteorological Station, Hull Street. These changing sites are an extra source of variation. The 52 year average (up to 1960) at the Kimberley police station is 400.8 mm and the 55 year average (up to 1960) at Kimberley-deBeers is 411.5 mm, while the 28 year average (up to 1960) at the Kimberley weather office is 427.3 mm (Climate of South Africa, pt. 9, 1965).

[b]$\bar{x}$ is the mean, σ the standard deviation, and CV% the coefficient of variation ($100 \cdot \sigma \cdot \bar{x}^{-1}$).

[c]In June, 1957, 105.9 mm of rain fell in 4 occurrences.

[d]In August, 1943, 117.6 mm of rain fell in 7 occurrences.

TABLE 2-4

EVAPORATION AT KIMBERLEY[a]

	J	F	M	A	M	J	J	A	S	O	N	D	Year
Class A Pan (mm)	363.2	289.5	246.4	185.4	127.0	99.0	129.5	185.4	254.0	337.8	373.3	378.4	2968.9
Symons Pan (mm)	266.7	220.9	187.9	137.1	88.9	71.1	88.9	132.0	170.1	241.3	266.7	281.9	2153.5

[a]These data cover a period of 4.5 years. Data are from Schulze (1965, Table 42).

at Kimberley are complex. For example during an average January situation, from 1900 hrs to 0700 hrs winds gradually veer from southwest to north-northeast and average less than 4 km per hour. However from 0700 hrs to 1900 hrs the winds gradually back from north-northeast to southwest and average greater than 7 km per hour. Eolian transport of sand and dust can be observed in the Lower Vaal Basin, but its role in soil erosion is limited and its effectiveness in modern landform sculpture, with the possible exception of the pans, is minimal. Strong winds most commonly occur along the convectional front of the thunderstorms and great volumes of dust are pushed upward, particularly from the disturbed soils in urban places and overgrazed or cropped areas. While the tracks of individual squall lines are erratic, a net westerly drift of the dust must be posited. At present large volumes of sand are not subject to eolian transport in the Lower Vaal Basin. Apart from minor areas along the Orange channel, there are no free sand dunes or sheets.

Landforms

The landscapes of the Lower Vaal Basin are dominated by broad, low-angle plains, bounded by river valleys or isolated clusters or chains of koppies. Landforms closely reflect the structure and lithology of the underlying bedrock. Upland areas, which typically have less than 100 m relief, are either exhumed relicts of the pre-Karoo erosional surface or buttes and mesas where a late-Karoo "dolerite" (diabase) sill forms a caprock for Dwyka shales beneath it. The plains are developed on less resistant shales or coincide with slightly undulating, Paleozoic plantation surfaces. Similarly the large-scale pool and riffle sequences in the rivers reflect the bedrock; pools often occur where the channel floor cuts across Dwyka shales, while riffles are most commonly found where channels intersect the more resistant rocks of the Pre-Karoo Surface. The valleys of the Harts, Vaal, and Orange are all partial re-excavations of pre-Karoo, glacially carved valleys. In addition, the most impressive landform in the basin, the Gaap Escarpment, is a pre-Karoo feature being exhumed with minimal alteration. Landscape profiles east of the Vaal are subdued by veneers of ancient dune sand -- the "Hutton Sands".

The Pre-Karoo Surface is extensively exposed on the Vaal-Harts interfluve, on the Nooitgedacht, Pniel, Rooipoort, and Spioenkop uplands to the south and east of the Vaal, on the chain of ridges west of the Vaal from Delportshoop to Douglas, and on the Gaap Plateau (Fig. 2-1). The Pre-Karoo Surface is composed of two different, Precambrian, rock groups -- the andesitic lavas of the Ventersdorp System and the clastic and biogenetic meta-sedimentaries of the Transvaal System (see also Chap. 3 below).

The volcanic terrains are found east of the Vaal-Harts valley, where Pre-Karoo Surface has low relief (seldom more than 30 m local relief per 10 km^2), and is characterized by gently undulating slope patterns. Slopes steeper than 8° are rare, and slopes less than 2° are common. The dominant profile elements in these planar areas are broad, low-angle, upper convexities which gently arch across the skyline (Fig. 2-2). Expectably, steeper slopes on volcanic bedrocks are found in high-relief terrains, particularly to the north of the Orange River valley, the constricted Riet River segment below Modderrivier, the gorge of the Orange River below the Vaal confluence, and the incised stretches of the Vaal River at Bushmanskloof (Fig. 2-3), between Barkly West and Delportshoop, and again between Warrenton and Windsorton. Relief in these areas may locally exceed 80 m, and straight mid-slopes (with recurrent angles of 12° to 17° and 22° to 27°) often constitute the dominant profile elements. Typically a much shorter convexity is found upslope. The concave

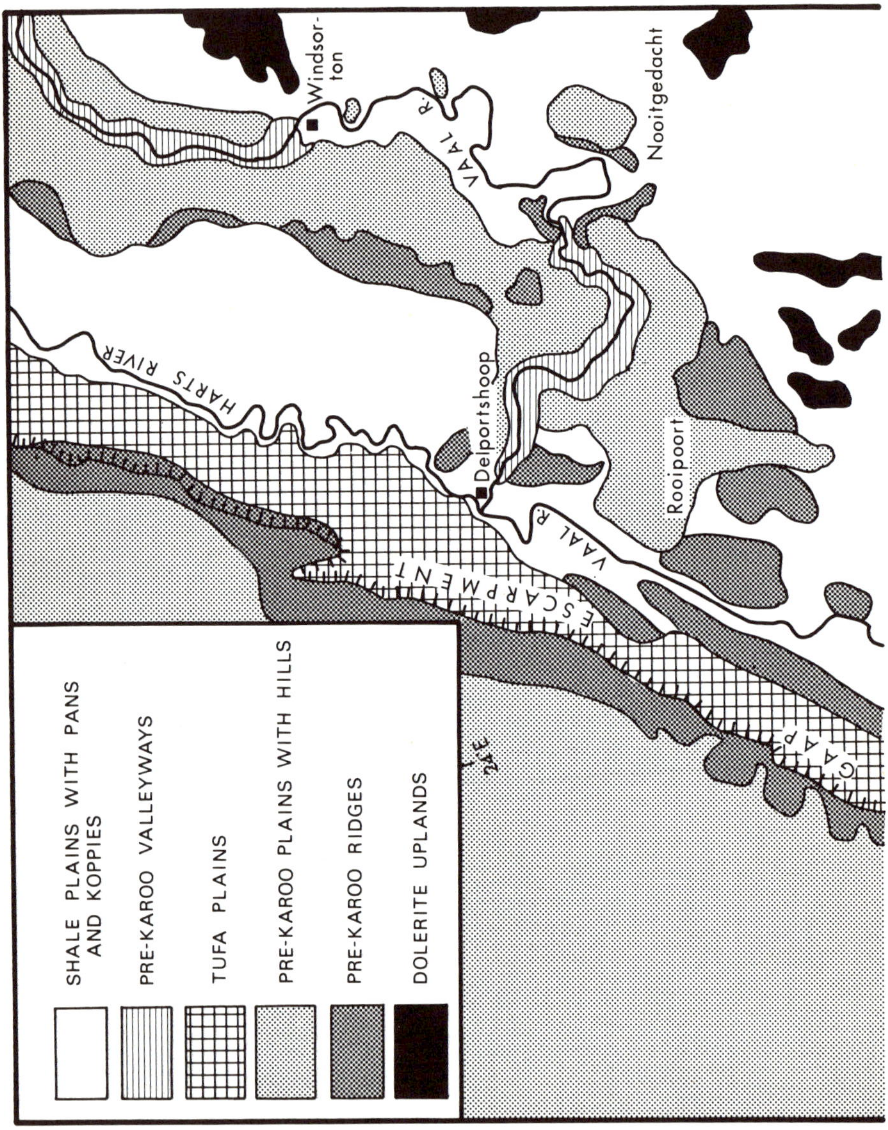

Windsorton
Nooitgedacht
VAAL R.
HARTS RIVER
Delportshoop
Rooipoort
VAAL R.
ESCARPMENT
GAAP
24°E
SHALE PLAINS WITH PANS AND KOPPIES
PRE-KAROO VALLEYWAYS
TUFA PLAINS
PRE-KAROO PLAINS WITH HILLS
PRE-KAROO RIDGES
DOLERITE UPLANDS

Fig. 2-1. Landform elements in the Lower Vaal Basin.

Fig. 2-2. Andesite hillsides along the Vaal channelway, 4 km north of Windsorton; in the foreground two lineaments cross.

Fig. 2-3. Andesite upland on the farm Rooipoort (background) at the head of the Bushmanskloof Gorge; in the foreground is a colluvial footslope merging with Riverton Formation alluvia.

footslopes are sometimes poorly developed due to preservation of Karoo sediments or extensive veneers of now-vegetated dune sands. The Ventersdorp volcanics have been severely fractured, and joints are closely spaced. Consequently weathered volcanic terrains present surfaces of subrounded, polygonal blocks measuring 0.3 to 2.0 m^3 (Fig. 3-14). However, in areas recently exhumed from the Karoo covermass, erosional records of late-Paleozoic glaciation are often preserved in detail. These include striations, crescentic gouges, and roches moutonées. Nearly all the volcanic terrains show some degree of glacial streamlining, and shallow, glacially scoured valleys trace the structural lineaments of the Pre-Karoo Surface.

The subhorizontal meta-sedimentaries of the Transvaal System (thinly bedded limestones, dolomites, shales, and quartzites) predominantly outcrop to the west of the Vaal-Harts valley. These terrains comprise the Gaap Plateau and Escarpment (Fig. 2-4), a low range of NE to SW - trending ridges along the Vaal, and a stretch of intervening plains. The escarpment crest tends to be gently undulating (at up to 2° parallel to the scarp), maintaining an average elevation near 1200 m. While the escarpment owes its origin to extensive pre-Karoo faulting and erosion, it received further sculpture during the Paleozoic glaciation. Subsequently it was buried by Karoo sediments. Depending on footslope conformation, (Fig. 2-5), the east-facing scarp is 60 to 100 m high, standing 100 to 200 m above the plains below. The plains extend 4 to 7 km eastward to the ridges that parallel the Vaal River. Standing only 30 to 90 m above the western plains, these NE-SW ridges rise 100 to 150 m above the nearby river channel.[3] Slope profiles in the NE-SW

---

[3]The crests of NE-SW ridges descend southwestwards at nearly the same rate as the modern Vaal channel, providing an interesting, but obscure symmetry.

Fig. 2-4. Gaap Escarpment north of Campbell; a small tufa carapace is in the left foreground.

Fig. 2-5. Gaap Escarpment footslope at Grootkloof; the resistant limestone beds are 2 to 4 m thick; in the background is the ULCO factory.

ridges exhibit well-developed upper convexities in macro-scale (with 12° to 16° steep segments common), but in detail they are often stair-cased with the outcrop of each bed forming a small facet. These ridges commonly present a surface of irregular, flagstone rubble, while individual limestone beds usually reveal an assortment of surficial, micro-karst features, particularly fluting and lapiés. The NE-SW ridges exhibit subtle glacial streamlining (drumlinoidal), while the plains in front of the escarpment are frequently underlain by glacigenic Dwyka shales.

After the late-Cenozoic removal of the Dwyka sediments that buried the Gaap Escarpment, calcium carbonate-charged surface and ground waters draining the karstic plateau deposited discrete bodies of limestone tufa along the scarp front and veneers of tufa on the fringing plains (Fig. 2-4). These tufas form conical carapaces or fan-shaped aprons of irregularly-bedded calcite flowstones. Each tufa body is the depositional record of a surface or subsurface drainage area on the plateau. Six major phases of tufa deposition have been recognized at several sites along the escarpment (Butzer, et al., 1978). Tufa aggradations signal relatively humid paleo-environments. Periods of tufa aggradation were followed by intervals of quiescence when the flowstones were significantly weathered and eroded. A solution-cavity filling in one of the oldest tufa units produced the first Australopithecine fossil, discovered in 1924 at Taung, 70 km north of Warrenton (see Peabody, 1954; Butzer, 1974a). In addition, two later tufa aggradations extend far beyond the scarp face to interdigitate with ancient Vaal and Harts River deposits. Flowstone aggradation closely followed the extant topography, so that along the escarpment crest, tufas are reposed in cliffs at 50° to 80°, while the surfaces of the larger carapaces are inclined at 10° to 15°; finally, the surface angle of the tufa bodies on the plains between the escarpment and the Vaal and Harts Rivers decreases from 3° to less than 1°. Although an unusually heavy rainfall or protracted rainy period on top of the plateau may lead to an equally brief period of flowstone deposition along the scarpface, no net tufa aggradation is occurring at present. Freshly deposited flowstones are eroded and weathered away before the next unusual magnitude event.

Away from the Pre-Karoo Surface most slopes are developed on subhorizontal Dwyka tillites and shales of late-Paleozoic age. Because the Dwyka rocks are weakly resistant to both mechanical and chemical weathering, the resulting landscapes are broad, low-angle (<2°) plains with straight to slightly concave profiles. These plains are interrupted where a sill of more resistant, late-Karoo "dolerite" (diabase) caps the shales, producing a mesaform koppie (Figs. 2-6 and 2-7). Typically, the koppies rise 90 to 100 m above the surrounding plains, but there may also be a few conical hillocks with a relief 20 to 30 m. Capping dolerite sills are 20 to 60 m thick. Koppie slopes are often smoothly concave in profile, with 33° to 37° angles near the top, 21° to 26° in the middle, and 5° to 8° at the base. Dolerite weathers spheroidally into subrounded, rhomboidal blocks of cobble to boulder size which litter the koppie slopes. It appears such fragments gradually slide and intermittently roll down these debris slopes, weathering along the way. Few dolerite chunks are found beyond the foot of the debris slope, indicating that most have entirely weathered away before this point is reached. Fresh shale or dolerite is commonly found less than one meter below the surface of the debris slope.[4]

---

[4]Interestingly, the upper slopes of the dolerite koppies are consistently 1° to 4° steeper on their south and southeast sides. This observational regularity may simply reflect the veneers of Hutton Sands often embanked on their northwest sides. A microclimatic factor may also be involved. It can be argued that south and southeast slopes are cooler and receive less insolation, so that chemical weathering would be reduced here, producing coarser-textured surface debris and a steeper slope angle.

Fig. 2-6. Dolerite koppies near Plooysburg.

Fig. 2-7. Dolerite koppie east of Douglas; footslope in foreground is heavily overgrazed.

The calcretes, produced by chemical weathering of the volcanics and the shales, comprise another caprock unit in the shale plain terrains of the Lower Vaal Basin. Generally confined to slopes of less than 4°, such calcretes extend from the very base of the koppies across the shale plains to form resistant crusts that protect the pediment-form plains from further denudation.[5] Precipitated in the shallow weathering zone from throughflow and soil waters, the calcretes mainly are a white, cryptocrystalline to crystalline calcite with additional silica cements, unweathered shale fragments, and quartz sands. There may also be weathered shale and dolerite debris. Since these wide, calcrete-coated, pediment-like plains extend smoothly downslope to the ancient terrace levels of the Vaal and its tributaries, it is clear that most of these calcretes are fossil in the modern landscape, so that the calcreted shale plains cannot be regarded as the product of contemporary slope processes. Even though this shale plain and dolerite koppie terrain is found in a semiarid context adequate sources of surficial sediments are not available to produce large alluvial fans. Intact bedrock is always present near the surface.

No systematic studies of modern slope processes have been undertaken in Griqualand West. Consequently, contemporary processes must remain largely a matter of conjecture.[6] However, the net water deficit throughout the year, the discontinuous vegetation mat, and the absence of frost-weathered sediments allow some reasonable speculations. With little water in the soil and rock-weathering zone, frost action and throughflow are severely restricted,

---

[5]The steepest, calcrete-covered slope this observer has recorded is 8°, at the Brakfontein Diggings, 45 km SW of the Orange-Vaal confluence.

[6]The high proportion of relict slopes in the region severly reduces the chances of determining process--form relationships.

while mass movements are limited to creep on the steeper slopes and rockfall from rare cliffs. The limited vegetation cover and high intensity, thunderstorm rainfall suggest that raindrop impact and surface wash, both channeled and unchanneled, are the most important means of erosion and overland sediment movement. Yet accumulations of slope sediments are only found in the valleys amid koppie clusters, and remain very thin (<50 cm) on the surrounding plains. Slope development patterns seem to systematically vary with bedrock structure and weathering characteristics. Exposures of pre-Karoo rocks usually reveal Paleozoic slopes that have suffered only minimal alteration since their exhumation. However, with either dolerite or calcrete acting as a caprock, the slopes on the shale terrains probably formed by backwearing and with only secondary downwearing. The calcretes armor the shale plains against further alteration.

Throughout the Orange and Vaal systems, river channels are incised below the adjoining plains, and the modern flood zones are scribed below a short sequence of higher terraces (see Butzer, 1971; van Zindern Bakker and Butzer, 1974). In the Lower Vaal Basin the river channels are often confined to valleys of the Pre-Karoo Surface, and long terrace sequences with much relict alluvium are best preserved here. Situated in these narrow bedrock valleys the channels have little opportunity for lateral channel movement (Fig. 2-8, 2-2). At most places the active channel is wider than the modern overbank deposits, and channel plans consistently reflect bedrock structure. Only a few valleyways of the lower Harts River are proportionately wide enough to accommodate important adjustments in channel plan. Sited on shales, the Harts channel is manifestly underfit, even within its mid-Pleistocene terraces (Fig. 2-9; Chap. 6 below). Where major bends in the other river courses do occur, they largely reflect pre-Karoo valley shapes, which, in

Fig. 2-8. Vaal channel 2 km above Gong-Gong; diggers debris lines the channel while the andesite slope on the opposite valleyside is dominated by glacially streamlined slopes.

Fig. 2-9. Harts valley at Spitskop.

turn, are largely determined by faults, master joints, and unconformities in the Precambrian bedrocks. In most cases what first appears to be modern floodplain terrains are in fact much older alluvial fills of later Pleistocene and Holocene age, comprising the Riverton Formation (see Butzer, Helgren, et al., 1973). Old channel traces are sometimes preserved on such floodplains, but these usually pertain to the aggradation of the Riverton Formation, and are only conveniently followed by modern floods. Modern overbank deposits of the Vaal River are grayish brown, clayey to sandy silts, while channel deposits are predominantly clayey to silty sands. Minor, local areas, particularly near ancient gravel exposures, have gravels in the modern load. Non-functional, coarse-cobble gravels frequently line the modern Vaal channel (Rietputs Formation, Member C; Chap. 5 below). These stabilized gravels are firmly compacted and lag concentrated, as well as frequently cemented at the top by ferric and manganese cements. They may bear Middle Pleistocene mammalian fossils. Modern sediments of the higher-volume Orange River are significantly coarser than those of the Vaal, being dominated by medium and coarse sands (Fig. 2-10).[7] Gravelly zones are not infrequent. The modern Riet River has a load of sandy silts and silty sands, while the Harts River deposits clayey silts and silty clays. Stream gauge records for the Lower Vaal Basin are presented in Table 2-5. The standard deviation computations provide an index of the erratic nature of the flows, which, of course, are an image of the variable magnitudes and durations of precipitation events in the basin.

---

[7]The Orange is actively building a silty sand spit across the Vaal mouth. This spit is breached by Vaal floods, only to be rapidly rebuilt by the Orange. The growth of this spit may be an artifact of recent flow management on the Vaal.

TABLE 2-5

SUMMARY OF STREAM GAUGE RECORDS IN OR NEAR THE LOWER VAAL BASIN

(Water Year Averages, Maximum and Minimum Yearly Flows)[a]

	Lat. S	Long. E	Record	Mean[b]	Standard Deviation	Maximum	Minimum
VAAL RIVER							
Bloemhof	27°39'	25°35'	1938 to 1969	229,415.	198,150.	972,290.	59,906.
Schoolplaats	28°06'	24°56'	1947 to 1972	160,689.	151,208.	615,776.	23,594.
Riverton	28°31'	24°42'	1909 to 1948	309,554.	308,058.	1,683,584.	40,637.
Riverton	28°31'	24°42'	1948 to 1970	155,364.	161,453.	630,738.	11,374.
Riverton	28°31'	24°42'	1909 to 1970	253,944.	273,496.	1,683,584.	11,374.
ORANGE RIVER							
Vluytjes Kraal	29°48'	24°26'	1948 to 1954	801,875.	375,270.	1,399,087.	127,479.
			1955 to 1964				
			1966 to 1970				
DRY HARTS RIVER							
Taung	27°34'	24°42'	1925 to 1946	1,024.	1,332.	5,839.	4.

HARTS RIVER							
Taung	27°34'	24°45'	1923 to 1967	3,715.	3,603.	16,655.	266.
			1968 to 1970				
Espagsdrift	27°54'	24°36'	1948 to 1956	7,126.	2,324.	10,547.	2,552.
			1959, 1962 to 1965				
			1966 to 1970				
MODDER RIVER							
Scholtzburg	29°01'	24°38'	1960 to 1970	6,966.	5,553.	15,780.	341.
RIET RIVER							
Aucampshoop	28°59'	24°12'	1953 to 1972	11,449.	11,962.	54,491.	200.

[a] Data are from Department of Water Affairs (1973). The accuracy of the gauges and quality of the observations vary greatly from site to site, and consequently these figures should only be viewed as "orders of magnitude". The data from Riverton are thought the most reliable.

[b] The units of measure are $10^4 m^3$.

Away from the major channels ephemeral streams gather rain-storm waters and direct them toward the rivers, although most of these ephemeral waters percolate into the ground before they reach permanent streams. On the Pre-Karoo Surface ephemeral channels tend to follow glacially-scoured furrows, or widened joint or fracture lines. But on the shale plains the ephemeral channels are shallow and discontinuous. Even high order reaches may be erased between storms through small-scale mass-wasting of the channel walls. These channels follow poorly defined surface screes of fluvially sorted materials. However, when these channelways reach the floodplain margins, they deeply incise the non-resistant flood-plain sediments (beyond 10 m) and form impressive erosional gulleys known as dongas (Fig. 2-11). Characterized by steep side walls and flat bottoms, the dongas start with steep headwalls near the flood-plain edge, as the channel base rapidly drops to near the local watertable. Dongas not only receive sediment pulses from the ephemeral stream, but the first flood waters of the through stream. On the floodplain the dongas often join to form dendritic networks before they enter the river channel. Certainly many dongas have deepened and extended their headwalls in response to the increased surface runoff and water table depression associated with agricul-tural settlement. However the stratigraphic sections revealed by many of the modern dongas show that their sites were also occupied by dongas during several previous phases of floodplain incision since the mid-Pleistocene. This suggests that donga location is primarily a function of the channeling of groundwater flow in depressions of the adjacent relatively impermeable bedrock.

The pans constitute the last important class of landforms in the Lower Vaal Basin. Pans are broad, shallow depressions with internal drainage. Here runoff waters accumulate following the

Fig. 2-10. Orange River below Hopetown, as it enters the Torquay Gorge.

Fig. 2-11. Dongas along the Riet River, below Schutsekama.

summer rains and form shallow, ephemeral lakes (Figs. 2-12 and 2-13). When the rains cease, the brackish and salt water lakes evaporate, leaving inhomogeneous, calcareous deposits in the saucer-shaped, pan basin. Surface sediments on the pans also fit within the generic class of calcretes, but opposed to the thinner calcretes on the shale plains, these materials are often massive and have important, highly variable, detrital components (see Butzer 1974b). As a result, pan sediments are sandy, loamy, and clayey marls, with interdigitating colluvial debris from the surrounding plains. They usually are inorganic calcites with mixtures of silicates, gypsums, and salts, but may sometimes include mollusca, diatoms, and algal secretions (Kent and Rogers, 1947). Unlike the similar playas of the southwestern United States and the sebkas of the Sahara, pans are of erosional origin. Dry season deflation and wet season, subaqueous weathering (corrosion) are apparently the important processes in net surface lowering. In response to differing bed-rock conformation and drainage basin size, pans are highly variable in size, shape, and internal relief. Small pans occupy structural depressions on planar segments of the Pre-Karoo Surface, but the large pans with long histories in Griqualand West are restricted to Dwyka shales. Like the rivers, the large pans are also incised below the surrounding plains and rimmed by sequences of higher terraces composed of ancient pan and freshwater lake deposits. The most ancient pan basins often are more than 10 km across, with modern pans more than 2 km wide.

Broad valleys sited in Dwyka shales often contain chains of pans. These flat, linear depressions are called vloers. While surface waters in the vloers may have flowed from pan to pan prior to this incision, pans are now only connected by limited groundwater flow. The origin of pans is not well understood and several principal factors appear to be involved (Geyser, 1950; Van Eeden, 1956; Butzer, 1974b). In Griqualand West the downvalley ends of many

Fig. 2-12. Pan on Brakdam (ca. 28°47'S; 24°23'E) dessicated, September 1970.

Fig. 2-13. Pan on Brakdam, full of water, March 1973.

pans often are marked by igneous thresholds--either dolerite outcrops or ridges of the Pre-Karoo Surface. Ponding of ground and surface waters behind these thresholds locally intensified shale weathering, so initiating pan formation. Dry season deflation was no doubt important in the excavation of the pan basins. Modern and relict, lunette dunes frequently occur to the lee (S and SE sides) of the pan basins. Sediment may also have been carried out on ungulate hoofs (Alison, 1899). The relict forms and deposits in the pan basins are an important part of the regional Quaternary record (see Butzer, Fock, *et al*., 1973).

Soils

Soils in the Lower Vaal Basin are intimately related to parent material and slope position. The most common soils are lithosols, which, by definition, exhibit minimal horizonation and humification. They are little more than rock weathering zones. For agricultural purposes, the soils on the alluvial sediments which fringe the rivers are most important. Other soils are found on bodies of former eolian sands now stabilized by vegetation, and at the footslopes of koppies and around pans, where the greater humidity of valley bottoms and colluvial accretions have conspired to produce heterogeneous soil patterns. The soils south of the Riet River have been mapped at 1:320,000 by van Rooyen (1972) (see also van Rooyen and Burger, 1974a; 1974b). A general classification for South African soils was presented by Van der Eyk, MacVicar and de Villiers in their study of the Tugela basin (1969). The formats of these authors are followed here. The soils of the Lower Vaal Basin are mapped in Fig. 2-14.

The most frequent surface veneers in the Lower Vaal Basin are the *calcrete* associations. Some calcretes are K-horizons of inceptisols, but in most settings the calcrete is relict, and the shallow, weathered, partly eolian and colluvial residues serving as the A-horizons are best viewed as lithosols. Calcretes are found on all types of bedrock, but they are most continuously developed on the shale and tillite plains and along the vloer valleys. Low slope angle appears to be a unifying characteristic. Calcretes are best developed on slopes of 2° or less. Lithologically the calcretes are variably white, cryptocrystalline to crystalline calcium carbonates with lesser admixtures of silica and other carbonate cements plus local detritus. Calcrete thicknesses range from 20 cm to greater than 2 m. The calcrete surface is often extremely hard to indurated, but hardness usually decreases with depth. Variability is a function of parent material lithology, availability of detritus from the surrounding slopes, groundwater relations and micro-environment history. A detailed study of calcrete structures and mineralogies in relation to these sources of variability has yet to be attempted. Some aspects of calcrete variability in the Lower Vaal Basin are presented in Table 2-6.

Both pedogenic and non-pedogenic, carbonate duricrusts must be included in the calcrete category in the Lower Vaal Basin. Genetic identifications are difficult by field inspection; many calcretes exhibit secondary re-crystalization and replacement, suggesting multiple phases of development, possibly by differing modes. Certainly the non-pedogenic, limestone-tufa surfaces on the plains in front of the Gaap Escarpment fit the general descriptions of calcrete, as do the ancient and modern evaporite deposits in the pan basins. Less obvious, non-pedogenic calcretes in the region have been produced by precipitation from throughflow waters and groundwaters, which are carbonate-charged from bedrock weathering upslope. These last calcretes are most obvious in relatively narrow bands in alluvial fills. The pedogenic examples have been produced by in situ weathering of the parent materials, particularly

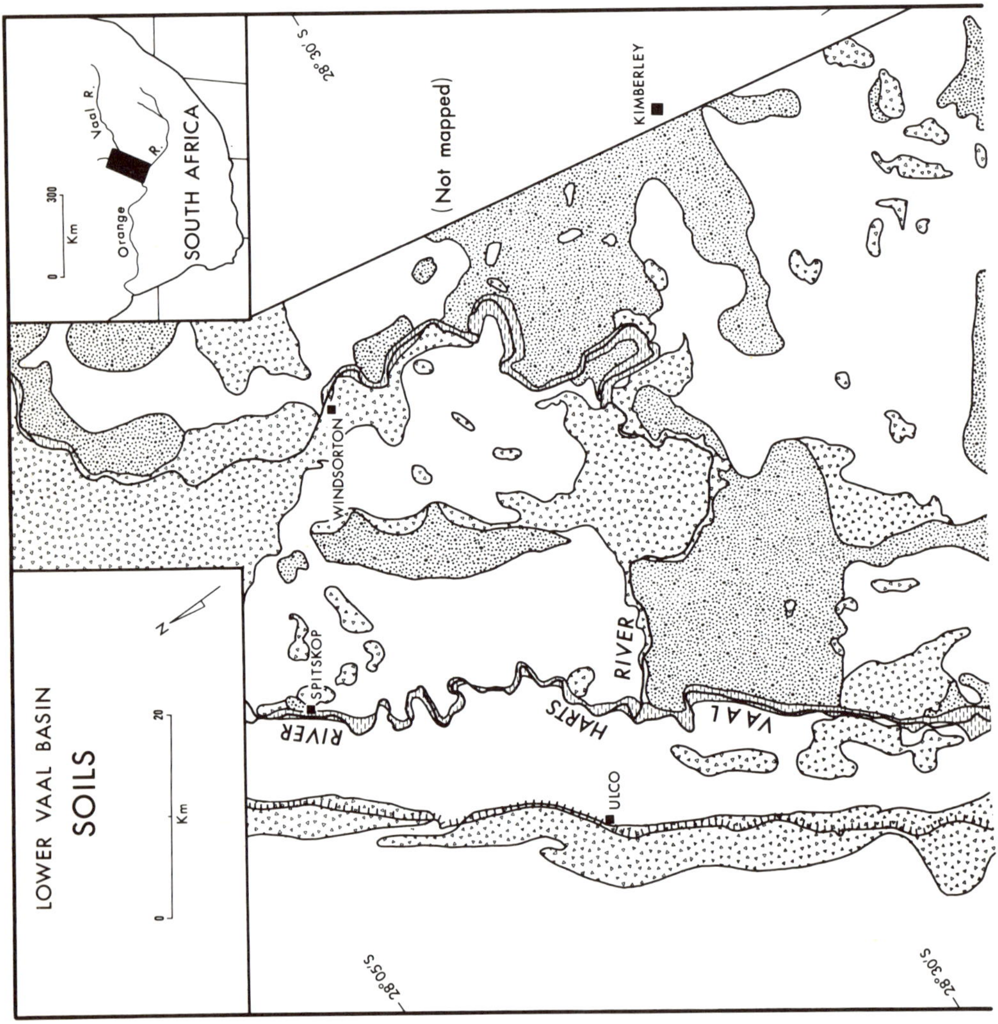
LOWER VAAL BASIN
SOILS
0
20
Km
N
0
300
Km
Orange
Vaal R.
R.
SOUTH AFRICA
28° 30' S
(Not mapped)
KIMBERLEY
WINDSORTON
SPITSKOP
RIVER
HARTS
RIVER
VAAL
ULCO
28°05'S
28°30'S

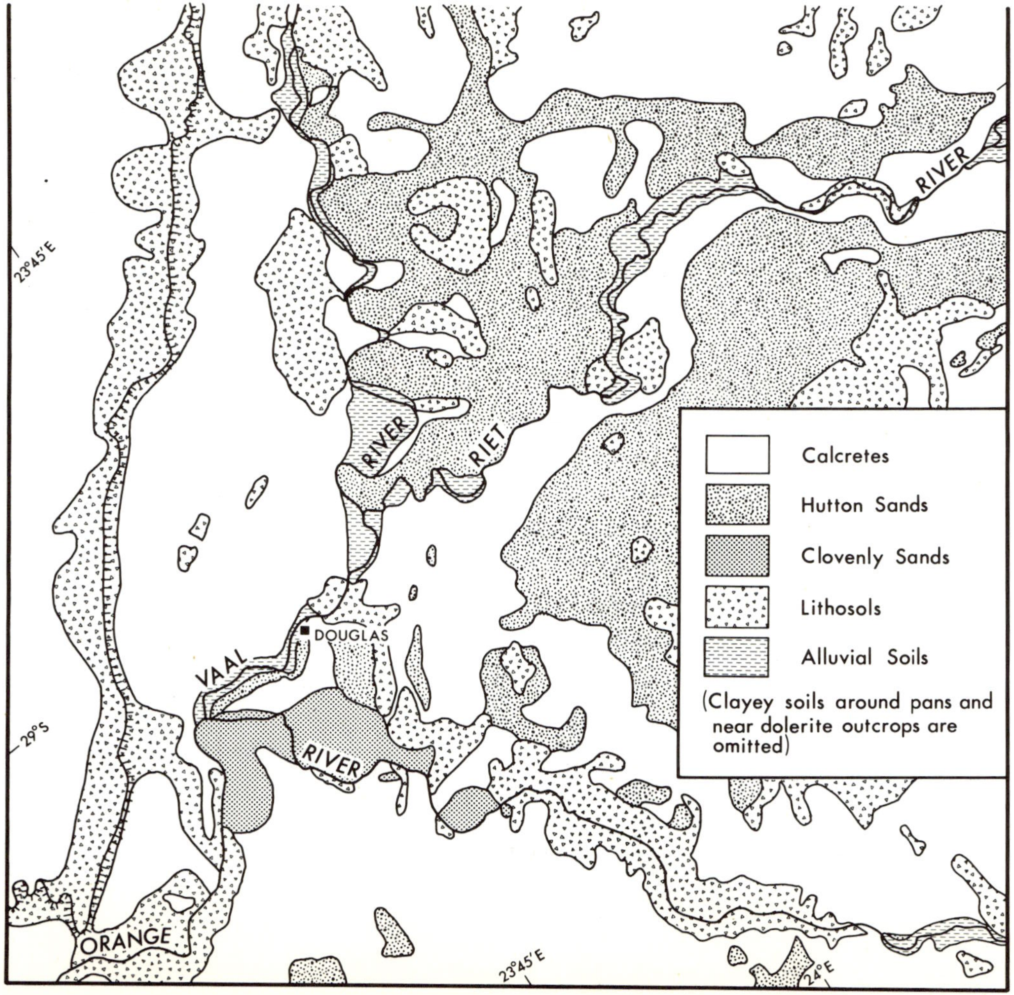

Fig. 2-14.
Generalized soil map of the Lower Vaal Basin

TABLE 2-6

VARIATIONS IN CALCRETE CHARACTERISTICS

Location	Surface age[b]	Thickness	Color	Hardness	% HCl soluble	SiO_2 amorphous	Texture of Insoluble Residue
Slangheuvel	Wedburg Terrace	3.5 m	N 10/0	moderately hard	83	7	sandy silt clay
De Bad	Wedburg Terrace	2.5 m	10 YR 7-8/3	extremely hard	83	11	clayey silt sand
Douglas	Rietputs A	1.6 m	2.5 Y 10/2	extremely hard	68	13	clayey silt sand
Koedoesberg	Rietputs A	2 m	2.5 Y 10/1	moderately hard	50	1	sand
Modderrivier	Rietputs A	1 m	10 YR 8-9/1	extremely hard	75	20	clayey sand silt
Modderrivier	Rietputs B	1 m	10 YR 10/1	moderately hard	61	11	clayey silt sand
De Kalk (on Riet)	Rietputs A	1.5 m	7.5 YR 9/1-2	moderately hard	70	10	clayey silt sand
Windsorton	Proksch Koppie Terrace	1.5 m	10 YR 9/2-3	extremely hard	70	2	sandy silt
Majeng	Rietputs A	2 m	10 YR 7-8/2-3	extremely hard	69	8	silty sand
Seskoppies (n of Holpan)	Older Gravels ?	2 m	10 YR 8-10/1-2	extremely hard	92	7	?
Thorngrove (w of Holpan)	Older Gravels ?	2 m	10 YR 10/1-2	moderately hard	79	20	?

[a]All samples occur on Dwyka shale or tillite bedrock. For laboratory procedures and textural criteria see Appendix A. For more on calcrete variability see the data in Netterberg (1969a, 1969b, 1974) and Goudie (1972, 1973). Texture descriptions follow Link (1966).

[b]This column indicates the age of the surface on which the calcrete is found, providing a maximum age estimate. These age relations are identified in the succeeding chapters below.

the shales and dolerites. Dissolved in weathering-zone waters, the carbonates and silica cements eventually rise toward the surface on capillary currents where precipitation occurs.[8] The processes involved in this "precipitation" are only poorly understood. When subjected to subaerial weathering, the calcretes may develop an extremely hard, laminated crust with a patinated and/or solution pitted surface. When eroded the calcretes rapidly abrade to subangular to subrounded pebbles and granules, which, in turn, may later be recemented. Calcretes are overlain typically by shallow, ochric A-horizons. Most commonly these A-horizons are pale brown (10 YR), loose apedal, sandy, silty, or clayey loams, which are less than 20 cm deep and have a basic to neutral reaction. Where a koppie, ephemeral channelway, or ancient eolian sandy body is nearby, the ochric, or sometimes mollic A-horizons are up to 45 cm deep and composed of reddish yellow (5 YR, 7.5 YR) to pale brown and brown (10 YR), neutral, loose, apedal, sandy loam to sand. In all cases the association of the A-horizon with the calcrete beneath it is not intimate. It appears that nearly all of the pedogenic calcretes in the Lower Vaal Basin are relict. Calcretes have not formed on lower Holocene alluvium along the rivers, despite the presence of appropriate, weatherable minerals and groundwater conditions.

The other class of lithosols in the Lower Vaal Basin are not calc-rich, but are typically composed of rubified, rock-weathering debris. They are characteristic on slopes steeper than 6° developed

---

[8]Pedogenic calcrete development is also closely related to the texture of the surrounding detrital materials. No pedogenic calcretes are found in medium to coarse sands, while medium and fine silts may be optimal. There may be an eolian (dustfall) component in calcrete genesis in this region, with deflation of the limestone and dolomite terrains of the Gaap Plateau a potentially important source.

on Karoo dolerites and pre-Karoo, volcanic and meta-sedimentary rocks. Here the shallow, ochric A-horizons are little but slightly humic, discolored rock waste. Characteristically it is loose, apedal, sandy loam to sand as a matrix to subrounded to angular pebbles of local bedrock. Variability in A-horizon development is a direct result of variations in bedrock type and in the local setting for bedrock weathering. On Ventersdorp andesite and Karoo dolerite the sand fractions are predominantly quartz with reddish, ferric coatings, while only limited rubefication occurs in sands on most sedimentary outcrops.

Narrow bands of soils on alluvial sediments adjoin the rivers in the Lower Vaal Basin. Typically they are fluvents with ochric A-horizons over stratified alluvium. However on channel edges and waterlogged sites vertisols are sometimes found. Depths of the A-horizons reach 60 cm, but 20 cm is more common. These soils are relatively youthful and have no B-horizons. The A-horizons are in the dark gray (10 YR) range, while the alluvial parent materials are grayish brown to brown (10 YR). Textures and structures are also relatively similar, being coarse to very coarse angular blocky, slightly hard, silty clay loams, clay loams, and silt loams. The pH's are neutral to slightly basic. At present these soils are being intensively irrigated, with general disregard for the potential problems of salt accumulation and localized water-logging. Also preserved on the floodplains are paleosols which developed at different stages during aggradation of the Riverton Formation. These range from vertisols through complex gleys to true pedogenic calcretes. Where these paleosols are found at or near the modern floodplain surface, the soil pattern is necessarily more complex. The calcareous examples probably are the most frequent paleosols.

Extensive eolian sands, now stabilized by vegetation, mantle much of the Lower Vaal Basin. Two varieties of sands are found -- "red" sands with Hutton form soils (the "Hutton Sands"), and "yellow" sands with Clovenly form soils (the "Clovenly Sands"; see also van Rooyen, 1972; van Rooyen and Burger, 1974b). The Clovenly Sands adjoin the modern Orange River floodplain, and relate to modern, small-scale eolian activity there. The largest area of active eolian slopeforms is found along the Orange near the Vaal confluence, while lesser areas extend up and down the Orange channel from there. Free sand surfaces are occupied by both merging barchans and low, longitudinal dunes. Depths of these sands derived from Orange River alluvium locally exceed 10 meters. Further away from the river, the undulating sand surfaces are vegetated and limited pedogenesis has occurred. These permeable, Clovenly form soils are characterized by shallow, ochric A-horizons over unweathered parent material. Both the A and "B"-horizons (actually C-horizons) are loose, apedal sands and loamy sands. Colors for the A-horizons may be as dark as brown (7.5 YR), while the parent materials are usually reddish yellow (7.5 YR) to pale brown (10 YR). The sand fractions are dominantly medium to fine-grade quartz, but notable proportions of highly weatherable minerals, such as pyroxenes and amphiboles, are also present. The weak horizonation and presence of weatherable minerals indicates the relative youthfulness of these soils. The Clovenly Sands rest on the modern Orange River floodplain, some of the Orange River's lower terraces, and on the slopes of the incised Orange River valley.

The widespread red sandy soils of the Hutton form extend far east and south from the Vaal and Harts Rivers and into the Riet River basin.[9] Deflated from ancient floodplains of the Vaal and Harts, these sands were subsequently remobilized several times between periods of vegetation colonization. Long, low, parallel ridges, representing the remains of subdued, large-scale longitudinal dunes, extend southeasterly from several reaches of the Vaal, most notably at the Harts River confluence (Fig. 2-15) and downstream of Schmidtsdrif. These sands spread across river terraces, areas of the Pre-Karoo Surface, and the calcrete-covered plains to bank against ridges and koppies of andesite and dolerite. When Hutton Sands rest on calcretes, as they commonly do, the calcrete surface often appears smoothed and patinated, perhaps through abrasion. Being highly permeable, the effectiveness of the hydraulic erosion processes on these red sands is severely limited. Sand depths sometimes exceed 12 m, but 1 to 5 m depths are most common. A typical Hutton profile shows a shallow, ochric A-horizon of slightly humic, apedal to weakly structured sand to sandy loam over a poorly defined B-horizon. The B-horizon is most commonly apedal, but may show vertical structuring, faint clay skins, and limited clay enrichment. The pH's are neutral to slightly basic throughout the profile. Clay contents are usually in the 4% to 12% range, but they may reach over 25% (van Rooyen, 1972). Clay content appears to be a function of both sorting and winnowing in the deposition of the parent material and the admixture

---

[9]Along the lower Harts River Hutton Sands are only found on the high shoulders of the valley. Along the Vaal Hutton Sands begin at the floodplain margin, and become progressively thinner and finer texture in an easterly direction. Traces of Hutton Sands are present as far east as Bloemfontein (van Rooyen, 1972).

Fig. 2-15. Hutton Sands terrain on Sydney, with the characteristic *Acacia giraffae* savanna.

of clays from local bedrock weathering. Clay proportions increase away from the rivers and are greater on steeper slopes, particularly on the windward side of dolerite koppies, where dolerite weathering debris has mixed with the original eolian component. The sand fractions contain no readily weatherable minerals, and are dominated by medium to fine grade, well-sorted and rounded quartz. The red color is produced by iron-oxide coatings on the sand grains. The Hutton soils are distinctive in the landscape of the Lower Vaal Basin. Not only do these sands mask the more angular shale plains and pre-Karoo topographies, but their red (2.5 YR) and particularly yellowish red (5 YR) colors are striking.

Two final soil types have limited distributions in the Lower Vaal Basin. First are the complex soils found on rocky footslopes surrounding dolerite exposures. These soils, the Shortlands form of van Rooyen (1972), have a weakly structured, sandy, ochric A-horizons over a blocky-structured, relatively clay-rich B-horizon. The thin A-horizons of reddish-brown (5 YR) with sandy loam to sandy clay loam reflect mixing with Hutton Sands, while the variably textured B-horizons of reddish brown to red clay loam reflect the accumulation of clays produced by dolerite weathering. Dolerite fragments are usually found throughout the profile, and zones of calcareous enrichment are common on low angle settings. Van Rooyen reasonably argues that these clay-rich B-horizons, being in relatively moist settings beneath the free-draining A-horizons, are the product of contemporary weathering processes. However the surficial mixing with Hutton-type sands equally suggests that they are in part paleosols.

The last soil type is only found in the fluctuation zone of the largest pans, and in ill-defined areas of ephemeral water accumulation on the calcreted shale plains. These very heterogeneous soils are the Katspruit and Willemsdal forms of van Rooyen (1972). In these periodically moist settings in situ weathering of Karoo sediments produces clay-, calc-, and salt-rich sediments, mixed with similar debris brought in by overland flow. In drier settings the shallow soils that result are simply loose, unstructured, ochric A-horizons of brown to pale brown (10 YR) silty clay loam over calcrete. But in moister settings there may be yellowish brown, gilgai-like, clay-rich soils with dry season desiccation cracks and subsurface gley horizons. Depths seldom exceed one meter. Some of these soils are found on pan terraces, and thus are clearly paleosols (see Butzer, Fock, et al., 1973), but the majority of occurrences clearly are contemporary. Figure 2-16 illustrates in profile the general association of soils and slopeforms in the Lower Vaal Basin.

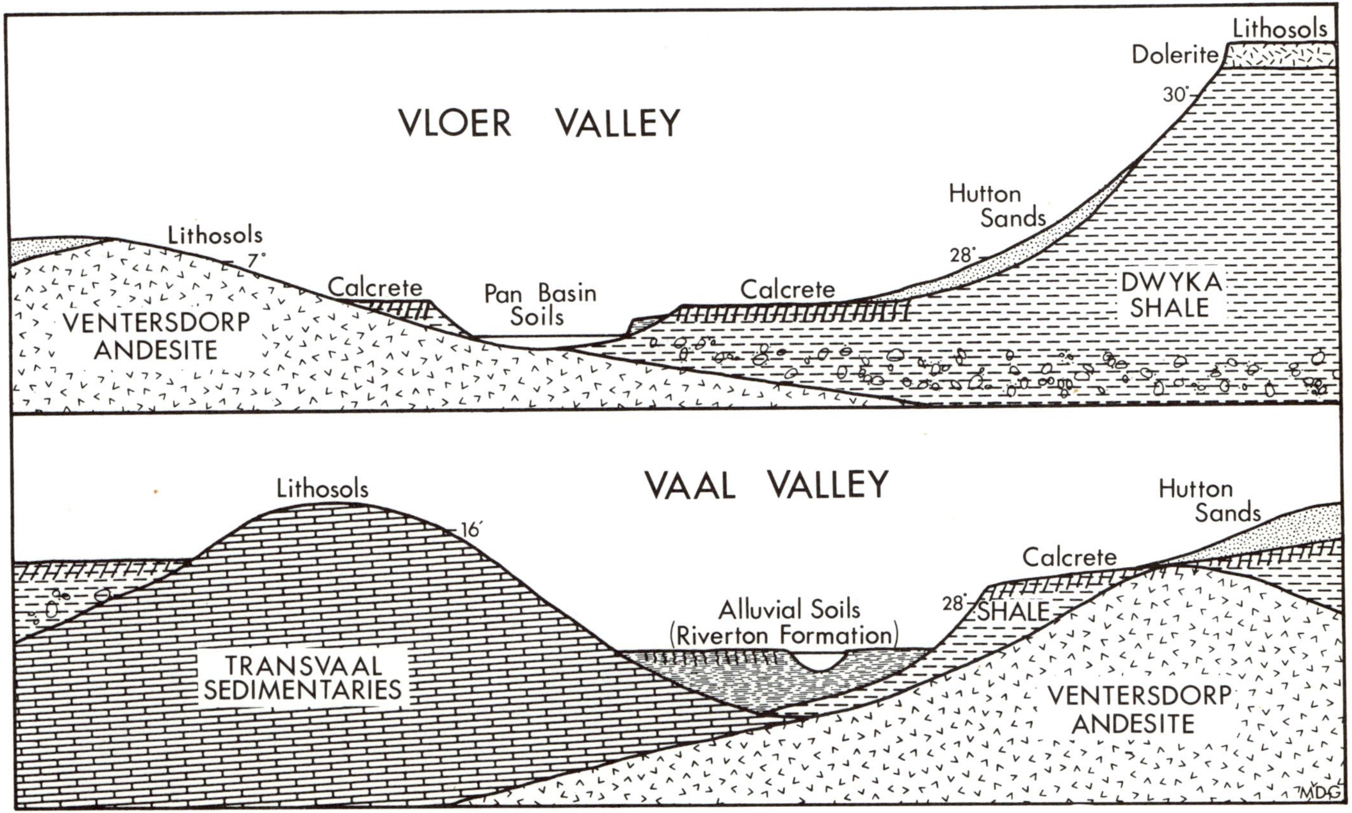

Fig. 2-16. Generalized soil and slopeform associations in the Lower Vaal Basin

Vegetation

Not surprisingly, floral patterns in Griqualand West are closely associated with soil distributions and slopeforms. However, on a broader scale, vegetation patterns in the Lower Vaal Basin also reflect the transitional position of this region between three major vegetation provinces of southern Africa. These provinces, according to Acocks (1953), are the bush savannas of the Kalahari Thornveld located to the north and west, the grasslands of the Dry Cymbopogon-Themeda Veld (Sour Grassveld) located to the east, and desert bush and bunch grass of the Karoo, located to the south. Consequently the landscapes east of the Vaal and south of Kimberley are dominated by karoid associations, while the landscapes west of the Vaal and north of Kimberley are dominated by Kalahari Thornveld associations. High Veld grasses fringe the eastern border. Acocks also argues that since the arrival of pastoral agriculturalists, the Karoo communities have moved dramatically northward, almost entirely displacing the Grassveld.[10] Overgrazing on all vegetation types has produced local areas of bare soil.

While the vegetation in the Lower Vaal Basin has been cataloged by Wilman (1946) and generally described by Acocks (1953) in his country-wide survey, the most informative data available are the descriptions of soil-vegetation relationships by van Rooyen (1972) and van Rooyen and Burger (1974a) and the detailed mapping by Bigalke and Leistner (1972) of Rooiport, a large farm northeast of Schmidtsdrif. The works of these authors are heavily borrowed in the following summary of typical vegetation and slopeform patterning.

---

[10] Pollen cores from the Alexandersfontein Pan area have been analyzed by L. Scott (1976 and unpublished). It appears that the vegetation mosaic as well as the dominant species near Kimberley have remained constant during most of the Holocene. Acocks (1953) worked on a large region with a limited data base, and many of his conclusions deserve re-evaluation.

Narrow gallery forests, seldom more than 30 m wide, fringe the rivers (Fig. 2-17). Along the Vaal and Orange the forests are taller, more continuous, and have a less drought-resistant composition than along the lesser streams. Tree heights reach 25 m, but heights in the 6 to 10 m range are more common. Salix capensis is the most common and impressive of these trees. Other important species are the thorny Acacia karoo, Zizyphus mucronata, and Rhus lancea. These species often form impenetrable thickets along the top of the river bank. Acacia karoo and Zizyphus mucronata, though sometimes only of shrub size, may be the most important species along lesser channels and in dongas. Shrubs, sometimes arboreal, such as Diospyros lycioides, Lycium hirsutum and L. echinatum are also common along the river banks. However few grasses are found here. On the modern and ancient floodplain sediments behind the forest fringe, the vegetation suddenly shifts to tuft grasses and low shrubs, seldom more than one meter high (Fig. 2-18). The shrubs are often riverbank arboreal species (including Acacia tortilis subsp. heteracantha), but Salsola glabrescens and Pentzia spp. are also present. The grasses include Sporobolus ioclados, Eragrostis lehmanniana, and Aristida congesta subsp. barbicollis. In general, this floodplain vegetation has been severely degraded in response to grazing pressures since agricultural settlement.

Beyond the floodplain margins, sands of eolian origin commonly mask the topography. On the "yellow" Clovenly Sands near the Orange River (Clovenly form soils), trees and bushes are infrequent. The important colonizers on free sand surfaces are the grasses Eragrostis pallens and Stipagrostis ambilis. On fully stabilized surfaces the most common grasses are Stipagrostis ciliata and S. obtusa. More diverse bush and tree associations, corresponding with the core of Acock's (1953) concept of a "Kalahari Thornveld",

Fig. 2-17. Dense gallery forest along the Vaal channel below Riverton.

Fig. 2-18. Grazing-stressed vegetation pattern along the Vaal at Willowbank; Wedburg Terrace is at left.

are found on the "red" sands (Hutton form soils) which spread east and south of the Vaal. Where the Hutton Sands are less than 1 m thick, a bush savanna is found with Rhigozum trichotomum and Tarchonanthus camphoratus as the dominant shrubs and Erogrostis lehmanniana and Aristida congesta subsp. barbicollis as the dominant grasses. R. trichotomum is usually used as an indicator of over-grazing and, perhaps with reduced grazing pressure, a perennial grass association might be expected to develop in these settings. As the Hutton Sands deepen, moisture availability increases at depth and a tree savanna results. On the plains the dominant trees of these impressive parklands are 8 to 15 m tall Acacia giraffae and A. haematoxylon (Fig. 2-15), while the important grasses (up to 1 m tall) include Eragrostis lehmanniana, E. pallens, Schmidtia pappophoroides, Stipagrostis spp., and Themeda triandra. When the red sands bank against a koppie and steeper slopes result, dense stands of Acacia mellifera subsp. detinens and A. tortilis subsp. heteracantha often become dominant. The correlation of this savanna parkland vegetation with deposits of thick, red, surficial sands is very close.

The vegetation on the calcrete lithosols has a karoid aspect, as it does on the shallow soils surrounding pans, and on the calcreted, shale-plains when Hutton Sands are absent. Here the uniformly-spaced, low (less than 2 m tall) bushes, tuft grasses, and intervening bare areas of pale to dark brown, shallow soil provide the monotonous landscape imagery for which the Karoo is famous (Figs. 8-6, 8-7). In detail, of course, there is much more variety to these plant associations than popular writers imply. The dominant shrubs are Tarchonanthus camphoratus, Grewia flava, Diospyros lycioides, Acacia mellifera subsp. detinens, Ehretia rigida, and stunted Acacia tortilis subsp. heteracantha. Also important are chamaephytes such as Pentzia calcarea, P. incana,

P. viridis, and Lycium oxycladum. The frequent grasses are Eragrostis lehmanniana, E. desvauxii, E. truncata, Aristida vestita, Fingerhuthia africana, and Cymbopogon plurinodis. Isolated trees, such as a stunted Zizyphus mucronata, Rhus lancea, or Acacia giraffae, may be found in less dry areas, such as a depression near a pan or spring, or along an ephemeral channelway. The floors of active pans are often devoid of vegetation. The karoo vegetation complex has a broad environmental tolerance, and the establishment of systematic relationships between soils, slopeforms, and plant species must await a region-wide phytogeographic survey.

The rocky slopes of dolerite and andesite koppies are clothed with an irregularly-spaced bush and low tree savanna. The bushes and highly palatable grasses freely protrude between the boulders and cobbles which litter the steep slopes. The dominant trees (up to 4 m tall) and bushes are Acacia mellifera subsp. detinens and A. tortilis subsp. heteracantha, while Boscia albitrunca, Zizyphus mucronata, Grewia flava, Euryops multifidus, Tarchonanthus camphoratus and Rhigozum obovatum are also important. The grasses, up to one meter tall, include Aristida vestita, Enneapogon scoparius, Eragrostis lehmanniana, and Stipagrostis uniplumis. At the base of the slopes where the sandy colluvia thicken, all the shrubs and trees listed above are found, but Zizyphus mucronata is particularly common. The more frequent grasses are Cymbopogon plurinodis, Sporbolus fimbriantus, Themeda triandra, and Eragrostis superba.

The final vegetation association, a monotonous Tarchonanthus bush savanna, is well developed along the Gaap escarpment and on the tufa plains to the west of the Vaal-Harts channel (Figs. 2-19, 2-20, 2-4, 2-5). The dominant shrub is Tarchonanthus camphoratus (1 to 4 m tall). The associated low trees and bushes are Rhus ciliata, R. pyroides, Grewia cana, and Olea africana. Where there is no grazing pressure, the grasses reach over a meter in height.

Fig. 2-19. Vegetation along the Gaap Escarpment footslope near Ulco, an ungrazed plot.

Fig. 2-20. Overgrazed Tarchonanthus savanna on a tufa surface north of Douglas.

The more frequent grasses include *Cenchrus ciliaris*, *Sporobolus fimbriatus*, *Cymbopogon plurinodis*, and *Aristida diffusa*. Along the base of the escarpment face, arboreal Acacias are often found, particularly *A. mellifera* subsp. *detinens* and *A. tortilis* subsp. *heteracantha*.

Beyond the Modern Landscape

On first observation the landscapes of the Lower Vaal Basin suggest landscapes found on desert margins throughout the world. Present are all the usual characteristics, such as the negative atmospheric and soil water balances, the open vegetation patterning, and the broad vistas of wide plains. However, on closer inspection, it is apparent from the shapes in the Pre-Karoo Surface, the paleosols on the plains and in the river valleys, and the high terraces on the rivers and the pans, that this region provides an extraordinary historical record. The geomorphic history is considered in the chapters below.

CHAPTER 3

PERSPECTIVES ON THE PRE-CENOZOIC GEOLOGY OF THE LOWER VAAL BASIN

> The present river system of Griqualand West and the adjoining Transvaal and Orange River Colony has proved to possess a history probably more complex than any other yet described from South Africa, and one that can be traced back clearly to a geological period considerably more remote.
>
> --A.L. du Toit (1910: 347)

Pre-Cenozoic Stratigraphy

The stratigraphic column in the Lower Vaal Basin is noteworthy for its long Precambrian sequence and record of late Paleozoic glaciation. At the base of the column are ancient granites and schists of the South African craton. These are succeeded by the metasediments and surficial volcanics of the Ventersdorp System and the fine-grained clastic and biogenic sedimentaries of the Transvaal System. The first Paleozoic events are recorded by the glaciation at the Carboniferous-Permian transition and verified by glacially scoured surfaces in Precambrian bedrocks and by the deposition of the tillites and shales that initiate the Karoo System. In the Jurassic these shales are intruded by basic magmas referable to the period of volcanic activity that marks the final stages of the Karoo System. The youngest pre-Cenozoic rocks are the late Mesozoic kimberlite pipes and fissures.

The diamond discoveries in Griqualand West precipitated several early inquiries into the geology of this region. Stow offered the first regional survey in 1874, and many more geologists recorded their passing reminiscences before the dreams of quick riches on the diamond fields faded away (see Wagner, 1914). The classic geological investigations are the regional surveys of A.L. du Toit (1907, 1908), who, working before the advent of aerial surveys, produced accurate maps and descriptions which remain invaluable today.[1] Also important are the observations by A.F. Williams (1932) on the geologic settings of the diamondiferous deposits in the region.

In this discussion the pre-Cenozoic stratigraphy is outlined first, followed by an analysis of pre-Cenozoic erosional forms. The pre-Cenozoic geology of the Lower Vaal Basin is portrayed in Figure 3-1.

Granites

The rarely-exposed, light pink to gray, usually foliated granites of the Lower Vaal Basin have minimal topographic significance. Small surface exposures are found along the north slopes of

---

[1]The "System-Series" nomenclature used here follows du Toit's original reports where both time and rock-stratigraphic units are implied. Du Toit's maps (at 1:238,000) have only recently begun to be replaced by a 1:125,000 series, of which only three, 2824 A -- Delportshoop, 2824 C -- Schmidtsdrif, and 2823 D -- Campbell, have been published at this writing. These provide much better data on the Transvaal System rocks, but most other areas were apparently mapped from air photos with limited ground survey, and many pre-Cenozoic outcrops are omitted.

the Orange River valley near Torquay, where they occupy the base of glacially scoured, anticlinal valleys. At most exposures weathering has intensively altered these meso- to macro-crystalline rocks, and recognition of their original mineralogy and textures is often difficult. Fresh exposures have many quartz and pegmatite veins. Presumably these rocks parallel the gneisses, schists, and foliated granites struck at depths greater than 300 m in the Kimberley diamond mines. The ancient erosional (?) surface on these Archean rocks at Kimberley exhibits over 600 m of local relief (Williams, 1932: 58), but this surface is not reflected in the modern landscape. The granites and associated, complex metasediments probably extend across all of the Lower Vaal Basin and broadly correlate with the Kraaipan Series of the western Transvaal (see van Eeden, 1963).[2]

Ventersdorp System

In the Lower Vaal Basin the Ventersdorp System is composed of the Pniel and the underlying Zoetlief Series.[3] Both series were defined by du Toit (1906, 1907), and his initial reports remain the

---

[2]A recent, exploratory borehole at Spitskop passed through 85 m of Dwyka sediments, and 983 m of Ventersdorp materials before basal chlorite schists were struck. Clearly the surface of the granite-schists has considerable relief.

[3]Whether the Zoetlief Series should be correlated with the Dominion Reef "System" at the base of the Witwatersrand System or placed at the base of the Ventersdorp System (and thus above the Witwatersrand sediments) has been a long-standing problem in South African, Precambrian stratigraphy (see Haughton, 1969: 125-126), which should be resolved soon by radiometric assays. While the Zoetlief Series underlies the Pniel Series unconformably, the two series are lithologically compatible and it seems unlikely that the long period of Witwatersrand sedimentation could have intervened. Winter (1976) correlates the Pniel Series with the Allanridge Andesite Formation in the Ventersdorp succession of the Orange Free State.

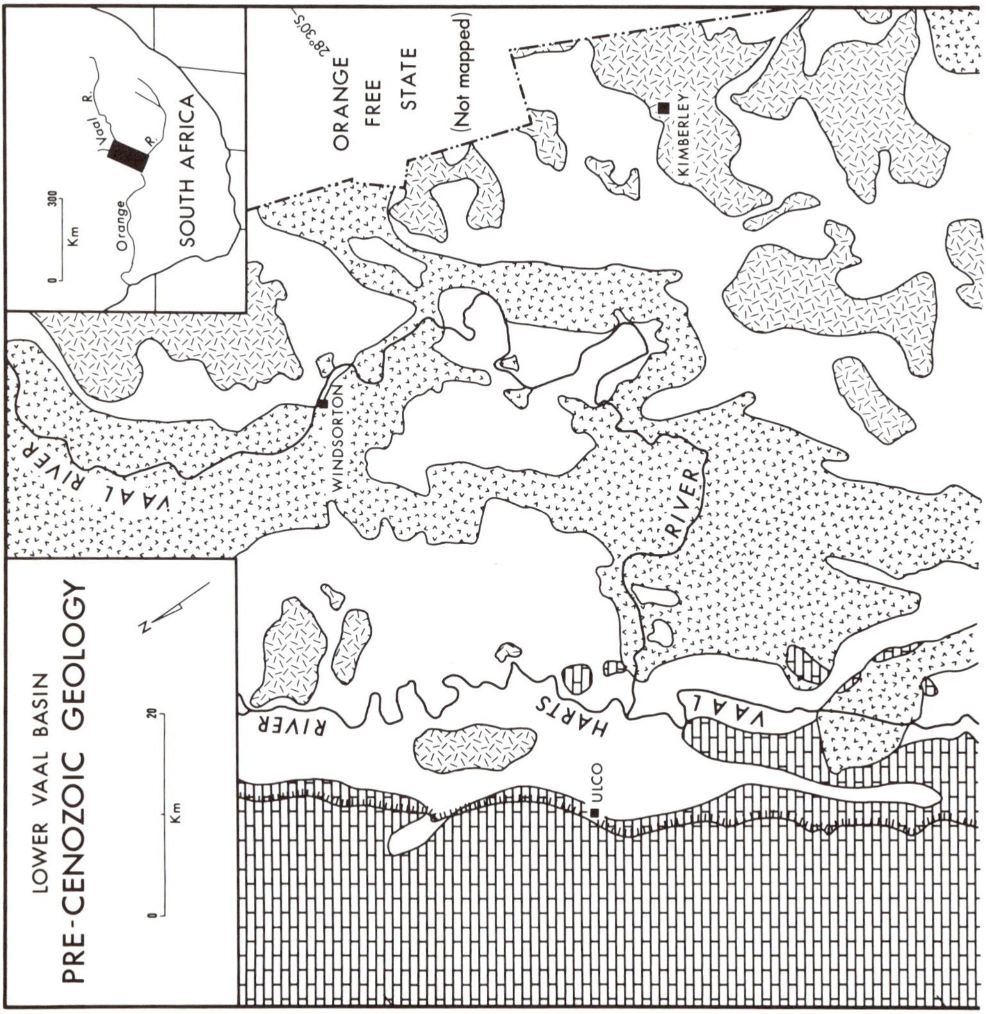
LOWER VAAL BASIN
PRE-CENOZOIC GEOLOGY
0
20
Km
0
300
Km
Orange
Vaal R.
R.
SOUTH AFRICA
28°30'S
ORANGE
FREE
STATE
(Not mapped)
KIMBERLEY
WINDSORTON
VAAL RIVER
RIVER
HARTS
RIVER
VAAL
ULCO

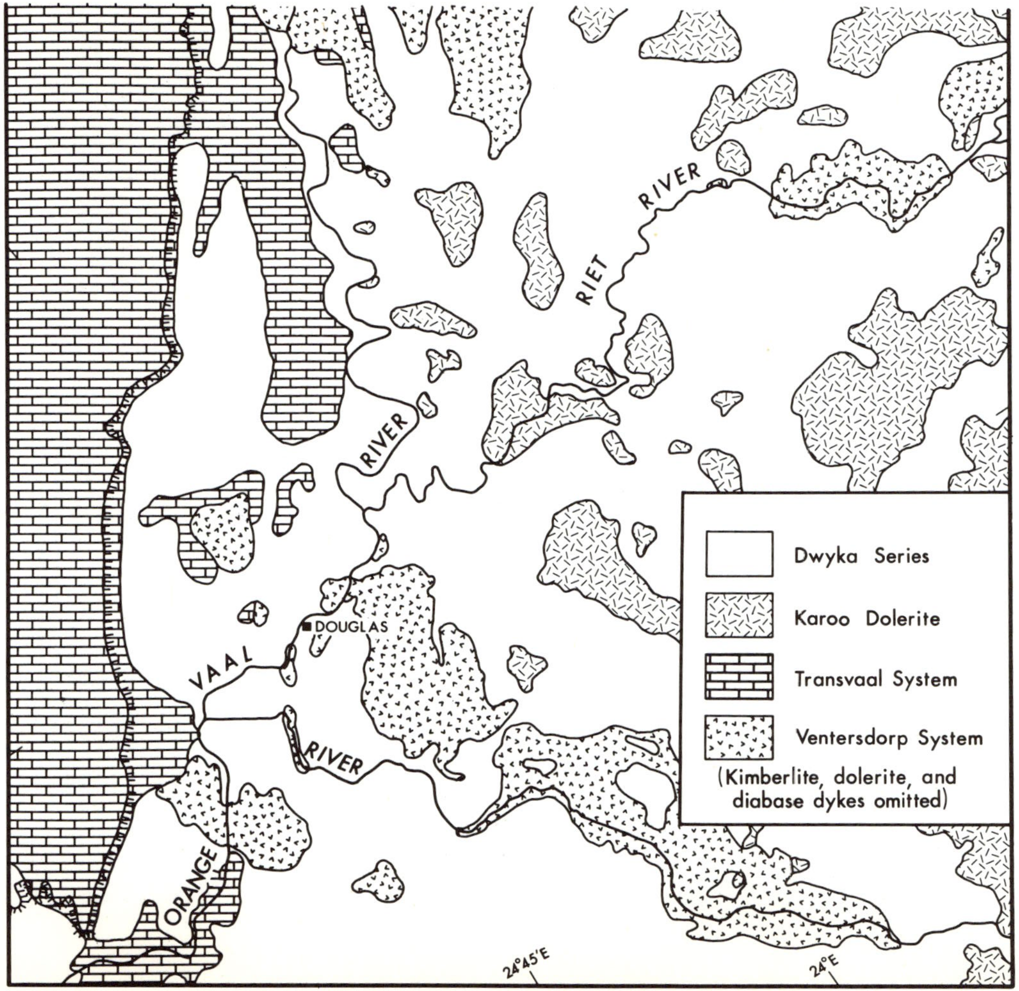

Fig. 3-1.
Generalized pre-Cenozoic geology of the Lower Vaal Basin

best systematic descriptions available. The thickness of the Ventersdorp System in the Lower Vaal Basin exceeds 100 meters. A uranium/lead date of 2300 $\pm$ 100 m.y. has been obtained on zircons from a quartz prophyry near the top of the Ventersdorp System (Van Niekerk and Berger, 1964; sample from outside the region).

Zoetlief rocks are only exposed along the Riet River, near the Modder River confluence. There they protrude through Dwyka shales and form five, small, dome-shaped koppies. Stratified outcrops occur in the Riet River gorge nearby. These exposures include rhyolitic quartz porphyries, rhyolites, and quartzites. As exposed in the Kimberley mines, the Zoetlief Series consists of a basal, sedimentary stage of conglomerates and quartzites, a middle stage of acidic, porphyrtic volcanics, and an upper stage of arkoses and quartzites (see Williams, 1932: 67-68). Only upper segments of this sequence are exposed near the Riet River.

The Pniel Series consists of a basal stage dominated by arenaceous metasediments and an upper stage of andesitic lavas. Localized tuffs and brecciated lavas occur in both stages. The little-exposed sedimentaries are clearly stratified, but bedding planes and directions for the complexly-structured andesite flows are difficult to determine. A systematic, region-wide stratigraphy applicable to the Ventersdorp System in the Lower Vaal Basin has not yet been developed.

Although the basal quartzites and sandstones have a total thickness in excess of 150 m (and in some areas they may be as thick as the capping lavas), they are not widely exposed. Exposures are found along the Riet River gorge and in a long, discontinuous band across the uplands running from Hopetown past Douglas. This last set of exposures is along the crest of an irregular, northwest-trending, anticlinal structure, here named the Spioenkop Anticline. In exposure these arenaceous strata typically are show gentle dips, reflecting the wider structural conformation. The most common rocks are light gray orthoquartzites, with some pinkish, iron-oxide stained, irregular quartz veins or localized quartzite pebbles. Rare arkoses and shales are also known. The topography in the quartzite terrains reflects the erosive activity of the Permian glaciers superimposed on even older structural shapes. No quartzite slopes are related to drainage lines developed on Dwyka sediments, which often fill the intervening valleys. Due to lithological constraints deep glacial striae are rarely found on the quartzites, but faint striae do occur. Some small-scale valleys are the product of glacial scour. Distinctive, large-scale landforms developed exclusively on such quartzites are low, cuestaform ridges and a 75 m high butte on the farm Clydesdale, south of Douglas. The Pniel Series quartzites are extremely resistant to modern weathering.

Du Toit (1907: 98) divided the upper volcanic stage of the Pniel Series into a basal zone of "normal diabases" and an upper zone of "porphyritic lavas, diabase, porphyritic diabase, quartz-porphyry and quartzite." While two koppies of quartzite and quartz-porphyry located near the Sydney-Rooipoort boundary apparently mark the upper part of the sequence in this region, systematic differentiation of the underlying volcanics has not yet proved possible. In general these lavas are more acidic than the term "diabase" connotes today. Here the whole sequence is generalized and labeled as the "Ventersdorp andesites", even though many of the amygdaloidal and porphyritic examples should probably be classed as dacites. No vents are known for these lavas. The presence of pillow structures and interbedded tuffs suggests a shallow, marine setting.

The Ventersdorp andesites are widely exposed. They occupy the highlands of the Vaal-Harts interfluve and the Nooitgedacht, Pniel, Sydney, Rooipoort (Fig. 2-3), and Spioenkop uplands. They are also exposed along most of the Vaal and Orange River valleys (Figs. 2-2, 2-8, 2-10) and in the Riet River gorge. Thus the andesites are exposed either as ridges and koppies overlooking valleys occupied by younger sediments, or are found along the incised river valleys. Slope profiles smoothly curve and ridgecrests are marked by irregular fields of boulders. Local valleys trace vectors of glacial scour superimposed on older structural fractures. Slopes are aligned with pre-Karoo drainage lines.

Small-scale features of glacial and glaciofluvial erosion can often be found on andesite surfaces. These include striations, grooves, and polished surfaces, as well as friction cracks (lunate fractures, crescentic gouges, etc.), potholes, and roches moutonnées (see Figs. 3-2 and 3-3).[4] Excellent examples of glacial pavements can be found along the Vaal River at Nooitgedacht and Riverton and along the Riet River at Driekops Eiland.

While the Ventersdorp andesites are consistently fine-grained, they occur with and without amygdules and with variable phenocryst development. Du Toit's "normal diabases" are non-amygdaloidal and vary in color from light green to light greenish-blue to dark blue and black. The amygdules, which are up to 10 cm in diameter, are composed dominantly of quartz (including jaspers, carnelians and agates), while calcite, feldspars and pyrites occur less commonly. The usually small phenocrysts are most often orthoclase, plagioclase,

---

[4] In general, the designation "roche moutonnée" has been misapplied in the Lower Vaal Basin. Embleton and King (1968: 152) point out that the term should refer to features which are smoothed and striated on the stoss side and plucked and quarried on the lee side. However, the "roches moutonnées" in Griqualand West are usually smoothed and striated on both sides, and thus should probably be classed as plastically-moulded features. Apart from joints crossing these pavements, they are as well preserved as any from the Pleistocene of North America.

Fig. 3-2. Striated andesite surface along the bed of the Riet River at Driekops Eiland; fine striations, gouges and joints superimposed by San (Bushman) rock art.

Fig. 3-3. Striated and gouged andesite surface along the Vaal River at Willowbank; two different striation vectors are present. The less obvious trend parallels the hammer handle.

and pyroxenes. Despite these variations in texture, the "matrices" of the amygdaloidal and prophyritic varieties, as well as the examples without amydules, are relatively uniform, propylitised andesites. This extensive alteration pre-dates Dwyka times. In most cases the ferromagnesian minerals have been almost completely replaced by chlorite and uralite pseudomorphs. Only the plagioclases show little change. Interbedded tuffs have been intensively rubified.

To investigate late Cenozoic weathering trends, a non-amygdaloidal example exposed near Winter's Rush was thin-sectioned across its 2 mm, reddish-brown weathering rind. This rind proved to be the product of selective weathering of the ferromagnesian minerals with chlorites and uralites largely to partly replaced by geothite. However, the plagioclases have remained largely unaffected. This unimpressive alteration probably is the product of most of Pleistocene time.

Transvaal System

In the Lower Vaal Basin the Transvaal System is represented by the Black Reef Series and the overlying Campbell Rand Series.[5] Black Reef rocks are exposed intermittently along the Vaal from Delportshoop to beyond the confluence with the Orange. Besides being exposed as mesaform ridges and koppies above the Black Reef, Campbell Rand rocks form the Gaap Plateau and are well-exposed along the scarp face (Figs. 2-4, 2-5). These two series are at least locally conformable and regionally dip northwestward at less

---

[5]Traditionally, the Campbell Rand Series has been called the Dolomite Series in the Transvaal. It is inappropriate to name a series after a rock type, and the term Campbell Rand Series has historical preeminence, being first used by Stow in 1874.

than 4°. However, the Transvaal System rocks have been faulted and fractured and, at any locality along the escarpment, dips may range from 0° to 6° to the west or northwest. Transvaal System rocks were not struck in the Kimberley mines.

The Black Reef Series, which has a total thickness of about 100 m, consists predominantly of light-colored quartzites and an upper unit of dark-colored flagstones. Thin interbeds of gray and green shales, tuff conglomerates, and carbonaceous sandstones are present. The Black Reef Series lies unconformable over the Ventersdorp volcanics (du Toit, 1908: 174-176), but the interbedded tuffs seem to suggest that no great temporal span separates them.

The Campbell Rand Series consists of a basal, transitional stage of fine-grained clastic and biogenic rocks and an upper stage of massive, dolomitic limestones, pure limestones and dolomites. Although the transitional stage has a thickness of more than 200 m, the upper carbonate stage may exceed 3000 m in total thickness before the western margin of the plateau is reached (Toens, 1966). The rocks exposed along the Gaap Escarpment largely belong to the transitional stage, with the first thick limestone beds occurring near the escarpment rim. The transitional stage has been described in three members and named the Schmidtsdrif Formation by Visser and Grobler (1972). With bed thicknesses seldom exceeding five meters, the irregularly cyclic basal member of the Schmidtsdrif Formation is composed of alternating dolomites, shales, siltstones, and quartzites. The middle member consists of green, gray and maroon silty shales with thin dolomite interbeds, while the upper member is composed of alternating dolomites, shales, and thin quartzose rocks. Seldom exposed along the scarp face, the two lower members are usually buried by talus or limestone tufa. The upper, dolomite-rich member and the capping massive limestones usually occupy the exposed scarp face. This capping unit includes well-preserved stromatolites, crypt-algal beds, and bioherms; an indirect, isotopic age of 2300 m.y. has been suggested (MacGregor, Truswell, and Eriksson, 1974; also Visser and Grobler, 1972). The depositional setting for the Transvaal System in the region apparently was a shallow epicontinental sea with a coastline retreating to the modern north.

Transvaal System rocks typically are exposed in staircased slope profiles, with each bed outcrop forming a distinctive slope facet. The shale beds are particularly non-durable. While glacial striations on the dolomite and limestone beds are uncommon, except on fresh outcrops, the flat-topped ridge crests often bear extensive, glacially-planed and gouged surfaces. The nearly flat Gaap Plateau is largely a karstic terrain with internal drainage. However, the large-scale sinkholes indicated by aerial reconnaissance are very shallow and obscured by pans or other sediment accumulations. Large underground cavern systems appear to be poorly developed. Shallow, surficial drainageways are present on the plateau surface, but they are usually dry. Following rainy periods water is ponded in pans on the plateau. At these times surface streams and springs pass over the escarpment, depositing the now-ephermeral, limestone tufa veneers.[6] The locations of springs and ancient tufas closely correlate with junctions of faults, master joints, and diabase dykes with the scarp face.[7] The limestone and dolomite exposures typically

---

[6]Rogers (1907: 79) provides an interesting account of tufa deposition in the nineteenth century. Apparently about 1 m of tufa was deposited shortly after 1879(?) on the farm Orange Grove (west of Schmidtsdrif). It was still present in 1906. The climate of South Africa during the last half of the nineteenth century would be worth systematic investigation.

[7]Diabase dykes along the scarp crest rise as linear ridges 10 to 25 m high. Groundwater ponding behind the dykes provides reliable aquifers, so that the dyke ridges have been loci for settlement.

exhibit small-scale solutional features, including pitting, small pipes, and lapiès ("elephant skin weathering"), which seem to be strongly controlled by crypt-algal structures.

Pre-Karoo Intrusions

A series of diabases, pegmatites, and quartz porphyries intrude the pre-Karoo rocks of the Lower Vaal Basin. Pegmatite dykes protrude through Ventersdorp volcanics on high Pniel (south of Barkly West) and along the Orange River at Marksdrif. A quartz porphyry dyke (the "Dikbosch Fracture", see below) trends across Ventersdorp terrain from Barkly West to north of Schmidtsdrif. Diabase dykes outcrop on some Ventersdorp surfaces, but they are most common along the crest of the Gaap Escarpment, where they form long ridges 20 to 100 m wide. These intrusions are obviously of different ages, both preceding and following the Transvaal System, but the timing of their emplacement and their relationships to each other are unclear. They are an important problem in the regional, Precambrian geology that could be solved by a detailed stratigraphic and chronometric investigation.

Karoo System

In the Lower Vaal Basin the Karoo System is represented by the Dwyka Series of tillites and shales and by late-Karoo intrusive volcanics.[8] The Dwyka Series, the basal member of the Karoo System, is the depositional product of late-Paleozoic glaciation in the region. Whether this glacial interval should be dated as late

---

[8]Obviously the Lower Vaal Basin was dominated by erosion through much of the Precambrian and early Paleozoic. Erosion of the Ventersdorp rocks probably provided an important fraction of the clasts for the early-Paleozoic Cape System.

Carboniferous or early Permian is clouded by the absence of diagnostic fossils. The Dwyka record for the whole Karoo basin has been summarized by Stratten (1968), and McLachlan and Anderson (1973) have described a fossil site near Douglas. The intrusive volcanics probably correlate in part with the late Triassic to early Jurassic Stormberg Series, which marks the termination of Karoo sedimentation. The Karoo volcanics of the region have not been studied in detail, but Walker and Poldevaart (1949) have offered a summary for the central Karoo basin, and McDougall (1963) and Fitch and Miller (1971) have provided a chronometric overview. In Lesotho Karoo volcanism and associated intrusive activity was initiated more than 180 m.y. ago, with activity continuing intermittently past 150 m.y. ago (Fitch and Miller, 1971). Apart from one small body of shales, on top of a dolerite koppie southwest of Modderrivier, which may be of Ecca age, no other Karoo System rocks are found in the lower basin. However, shales of the Ecca Series closely adjoin the eastern boundary.[9]

At the base of the Dwyka Series in the Lower Vaal Basin is a thin, extremely hard tillite which bears striations (first reported by Rogers and du Toit, 1909: 177; but generally neglected subsequently). Seldom more than 30 cm thick, this tillite is dark brown to black in color and commonly overlies Ventersdorp andesite, which is itself striated. This often-unstratified, somewhat arenaceous tillite is exposed along the Vaal-Harts interfluve, with good

---

[9]Fragments of later Karoo rocks and fossils collapsed from sidewalls of the kimberlite pipes in the region and have been recovered by mining. The contacting Karoo sedimentaries show distinctive pressure metamorphism (Hawthorne, 1975; Rowsell and de Swardt, 1976; Williams, 1932). Representatives of all the Karoo System (greater than 2000 m thick) were present in the Lower Vaal Basin at the time of the kimberlite emplacements (late Cretaceous).

exposures found near Holpan on the farm Rocklands (Fig. 3-4). This basal tillite is the same one found overlying the famous glacial pavements at Nooitgedacht, where it appears to have suffered the same plastic-moulding as the andesites (Figs. 3-5, 3-6). At Willowbank (north of Riverton) there are impressive flutings in the lower tillite which rise as staircased furrows with 20 to 70 cm of relief (Figs. 3-7, 3-8).[10] No pre-Dwyka regolith has been discovered. A second glacial period in Griqualand West is recorded by the regular array of often-stratified tillites (presumably aquatic, ice contact moraines) and dropstone beds (Fig. 3-9). These horizons have softer, bluish and greenish-gray matrices and are much more widely exposed, particularly along the Vaal below Schmidtsdrif and along the Orange River, both above and below the Vaal River confluence. This second tillite sequence grades gradually upward through dropstone zones into the black, gray, blue, and green shales all with rare, thin limestones and sandstones. The Dwyka series must exceed 400 m in total thickness in the Lower Vaal Basin, with this thickness taken up about equally by the tillites and shales. However, there are no comprehensive exposures and a detailed, systematic lithostratigraphy of the series has yet to be completed. Any of the tillites or shale beds may be found at the contact between the Dwyka series and the Pre-Karoo Surface beneath it.

The tillites consist of boulder to sand-grade particles set in fine-grained, argillaceous matrices. Only seldomly can the matrices be described as sandy. The shaley matrices contain few clay minerals, but are dominated by silt and clay-sized microclasts (Hamilton and Krinsley, 1967). The matrices are mixtures of angular to sub-rounded quartz with secondary feldspars (particularly plagioclases) and micas.

---

[10]This basal tillite is also nicely preserved on the west slope (lee-side relative to the second glaciation) of Amandelhoogte near Windsorton.

Fig. 3-4. Striated tillite over separately striated Ventersdorp surface, on the farm Rocklands near Holpan.

Fig. 3-5. Glacial pavings at Nooitgedacht; striated and jointed andesite mound (background) veneered by basal tillite. The largest tillite clasts have 30 to 40 cm long-axes.

Fig. 3-6. Basal tillite and striated andesite at Nooitgedacht.

Fig. 3-7. Striated andesite surface at Willowbank; tillite veneer in background.

Fig. 3-8. Fluted tillite at Willowbank; flutes have 30 to 70 cm of relief.

Fig. 3-9. Dropstone bed along the Vaal at De Bad.

Lithologies of the pebbles, cobbles, and boulders largely reflect the local, pre-glacial bedrock. In most cases the larger clasts are angular to rounded fragments of the Ventersdorp andesites, but in the glacial valleys along the Orange below Douglas, the Transvaal System limestones and dolomites are also common. The larger fragments, which may be two meters in length, are always locally derived, but a few pebbles and many clay and silt-sized microclasts are from further afield and may have lithologies no longer available in the Vaal and Orange River drainages. The gravel clasts from two tillites (at Nooitgedacht and Cyphex) and from one dropstone bed were morphometrically analyzed (Table 4-1). While it was found that average sizes, shapes and roundness for these samples were not significantly different, the dropstone sample was poorer sorted and contained a much lower proportion of striated and faceted stones

(1% to 16% and 26% respectively). Due to the large proportion of small to large stones in any one exposure, principal axis length averaged between five and seven centimeters. The shale matrices of the dropstone horizons are, in part, varved, and frequently show depressed and ruptured laminations.

Beginning above the tillites and partly interdigitating with the dropstone horizons are the horizontally-bedded Dwyka shales. They are most commonly dark gray, but blue and green beds also occur. Thin limestone interbeds are also known. Lithologically they are similar to the aquatic tillite matrices. Van Rooyen (1972) determined the chemical composition of three Dwyka shale samples (locations are not given). The samples averaged 50% SiO_2, 16% Al_2O_3, 5% Fe_2O_3, 6% CaO, 3% MgO, and 3% K_2O. A variety of marine fossils have been found in the Dwyka shales, including cephalopods, lamellibranchs (Phestia spp.), brachiopods (Attenuatella spp.), fishes, and the aquatic dinosaur Mesosaurus (McLachlan and Anderson, 1973). Fossil wood is also known. The surface of the shale is commonly calcreted, with irregular veins of calcrete penetrating the shale up to a depth of three meters. These calcrete veins, which perhaps follow "channels" of capillary currents, regularly contort the finely-bedded, soft and fissile shales.

A startling characteristic of the Dwyka Series in the Lower Vaal Basin is the near total absence of any associated sandy deposits. Thin, basal sandstones locally only occur between the Riet and Orange Rivers.[11] While the sequence quickly goes from terrestial to aquatic conditions and includes at least two glacial

---

[11]This area of sandy beds extends southwestwards from Modderrivier toward the Spioenkop Anticline, suggesting some sort of submarine channelway. A 20 m thick, 100 m wide, wedge of cross-bedded sands occurs below the south margin of the Gaap Escarpment (on the farm Naras). The upper tillites in this area carry a 10 m thick, dark red band, which is the product of abundant ironstone micro-clasts from the Asbestos Mountains.

episodes, there are no beach deposits, no sandy outwash (sandur) deposits, none of the thick typical terrestial glacio-fluvial deposits (i.e. kames, eskers, etc.),[12] and no local dune deposits (see Reading and Walker, 1966, for a more expectable set of facies). This presents problems in finding a modern or Pleistocene depositional analogue for the series. The basal tillite and probably the lowest parts of the softer tillite above it are terrestially-deposited ground moraines, but the sequence quickly becomes aquatic and probably marine before the appearance of the first dropstone horizon. A first approximation at an environmental model suggests deposition along a retreating, high-latitude, glacier front on a transgressing, low-energy coastline, which was backed by a foreland with irregular hills and glacially widened valleys.[13] The Dwyka series probably records only the last events in a long sequence of repeated glaciations. The landscapes of the Dwyka glaciation are considered further below.

The late-Karoo intrusives have traditionally been called "dolerites". Even though modern North American usage would label the examples in the Lower Vaal Basin as diabases, the term "dolerite" is continued here. The dolerite exposures in the Lower Vaal Basin have not been systematically studied, and du Toit (1907) and Williams (1932: 72-79) provide the best accounts. Most of the present exposures are largely referable to one, large sill which once covered nearly all the region. However, less-extensive emplacements have also occurred, possibly at much later dates. These include minor dykes, the largest of which are visible as

---

[12]L.N.J. Engelbrecht (1973) reports thin, fluvially sorted outwash deposits on the andesite uplands north and east of Barkly West.

[13]Modern and Pleistocene analogues may be present in some of the low relief, crystalline-rock coastlines of northeast Canada, Greenland, and Antarctica.

photo lineaments. Similarly there are an assortment of lesser sills, particularly along the Riet River near Schutsekama and Driekops Eiland. The region-wide sill has a maximum thickness of about 90 m, but 20 to 40 m is more common. This dolerite sheet is often horizontal but, as noted by du Toit (1907; 122-124), it generally parallels the Pre-Karoo Surface by 30 to 80 m, rising and falling with the major undulations of the surface. The general stratigraphic level of this intrusion is near the top of the upper Dwyka shales. On exposure the dolerites usually weather spheriodally (Fig. 2-7); columnar-jointed exposures are known, but rare. Contact metamorphics irregularly follow the sill margins; these are the "indurated shales" and "lydianites," popular raw materials in Stone Age times. However, in many cases the adjoining shales have suffered only minimal alteration.

Mineralogy of dolerites in the Lower Vaal Basin is variable. All examples are medium to fine-grained diabases, but the proportions of olivine, pyroxenes, and quartz vary. Williams (1932: 79) reports that the chemical composition of an "average" sill in the Kimberley area is about 48% SiO_2, 16% Al_2O_3, 5% Fe_2O_3, 6% FeO, 9% CaO, 6% MgO, and 3% Na_2O. A photomicrograph of a thin olivine diabase dyke from near Delportshoop is presented in Fig. 3-10. Clearly illustrated is the equigranular, diabasic texture with plagioclase laths penetrating anhedra of clinopyroxenes. The absence of alteration is significant relative to the Ventersdorp andesites.

Kimberlite

The last pre-Cenozoic events are marked by intrusions of Kimberlite pipes and fissures. Dates as late as 70 m.y. B.P. seem assured, while the earliest may have an age of 140 m.y. B.P. (J.B. Hawthorne, personal communication). Whereas the most famous

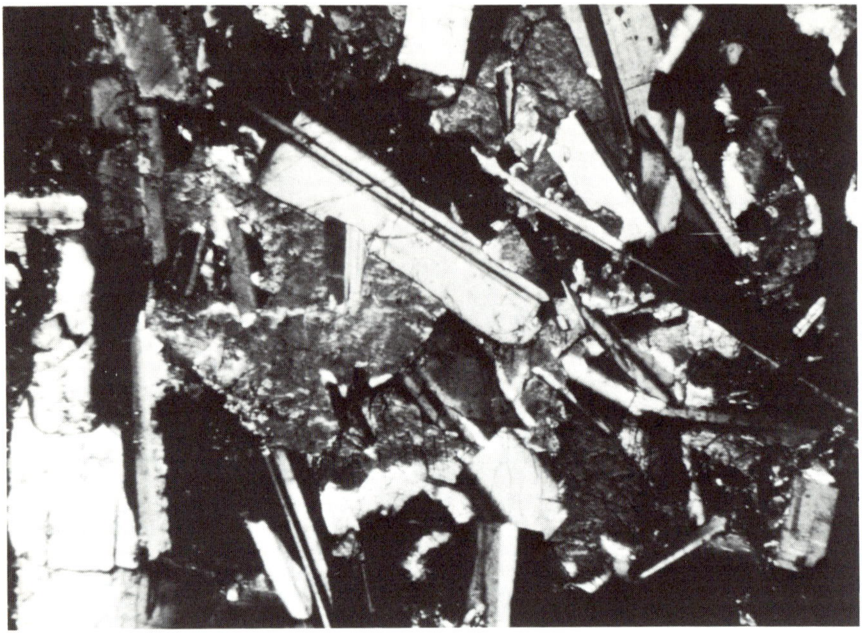

Fig. 3-10. Photomicrograph of a Karoo Dolerite; "fresh" diabasic texture is typical. Frame width is 2.0 mm.

exposures of this ultra-basic, olivine-rich rock are at Kimberley, lesser fissures radiate outwards in all directions (see Wagner, 1914). However, not all kimberlites are diamondiferous, and only a few exposures are marked by mines. The topographic significance of the kimberlites is limited. Where large pipes originally occurred, they were represented by low, dome-shaped koppies. Now the sites of these koppies are deep pits fringed by mining dumps. Most of the smaller fissures have no topographic expression at all. A line of small fissures marks the top of the Vaal-Harts interfluve. These kimberlites probably provided most of the diamonds found in the Vaal River gravel terraces. However, some of the diamonds in the alluvial gravels were brought into the region via late Paleozoic glacial or late-Cenozoic fluvial transport.[14]

The Pre-Karoo Surface

In Griqualand West, Dwyka sediments with nearby glacially eroded, Precambrian surfaces can be found almost continuously along the modern channels of the Vaal and Orange Rivers (Fig. 3-11). Similarly the Riet River flows near or on the Pre-Karoo Surface below its junction with the Modder River. Only the Harts River flows exclusively on Karoo sediments and volcanics above the Pre-Karoo Surface. Thus it is clear that the modern river valleys in the Lower Vaal Basin are re-excavations of valleys formed during or prior to the Permian glaciation. These characteristics of the landscapes in the Lower Vaal Basin were noted by du Toit in 1910. He labeled the great valley along the east and south margin of the Gaap plateau as the "Kaap Valley", and recognized it and several lesser valleys as primary routes of the Paleozoic glaciers. However, he explicitly recognized that the glaciers tended to follow pre-existing drainageways and that his glacial valleys also were pre-glacial valleys. Detailed field inspection and air photo analyses in this investigation have demonstrated the general accuracy of du Toit's identifications.

---

[14]Because of the secrecy surrounding both the legal and illegal sides of the diamond industry, accurate diamond provenience data is difficult to acquire. Most of the alluvial diamonds came from erosion of kimberlites in the region, but others were transported from the upper Vaal basin, from sites such as the Premier Mine. The Premier pipe near Pretoria is Precambrian, and such ancient kimberlites probably supplied diamonds to the Paleozoic tills. The author knows of one tillite outcrop that supposedly produced diamonds.

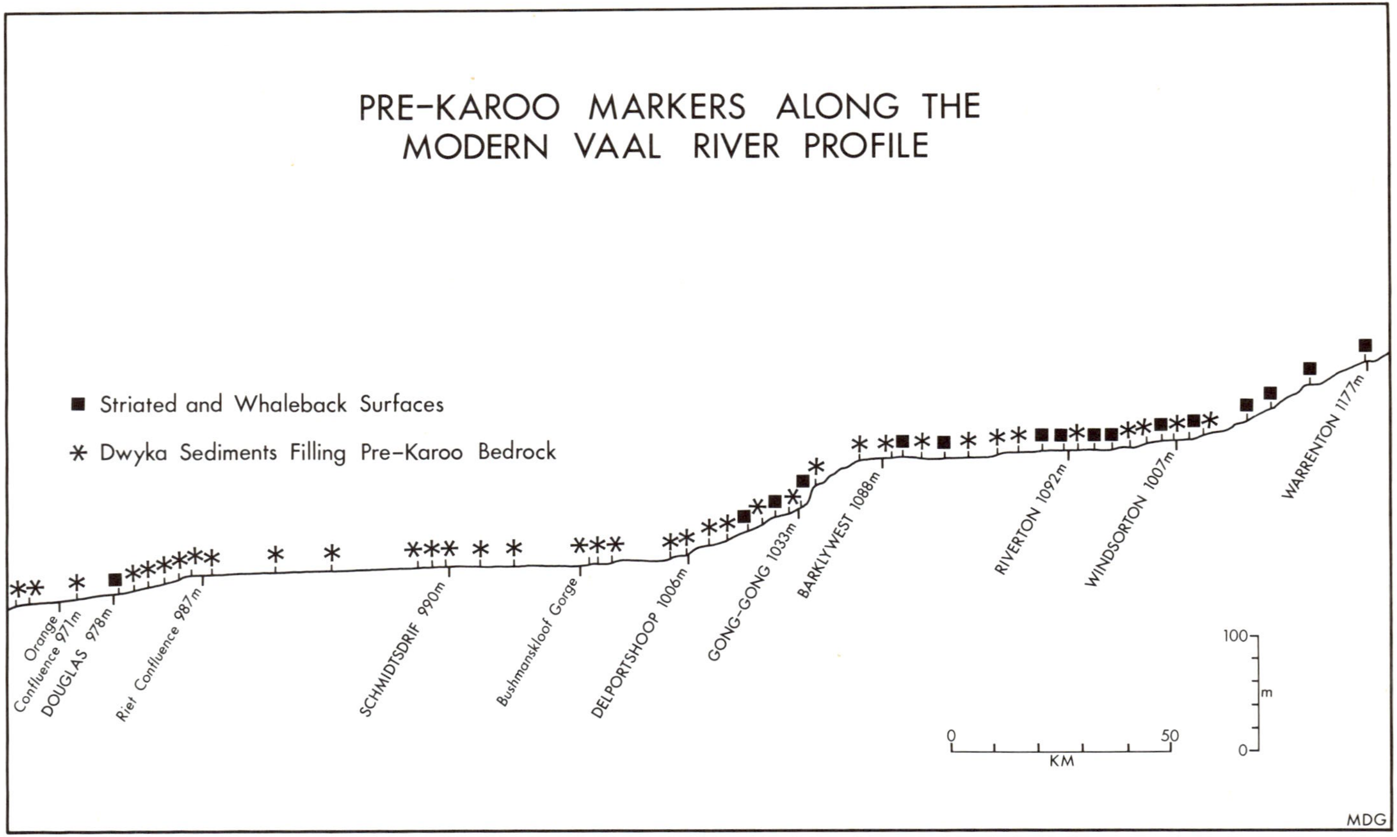

Fig. 3-11. Glacial markers along the modern Vaal River.

In evaluating the Dwyka glaciation it is necessary to use the Pleistocene glaciations as an analogue model. Estimates of Pleistocene glacial denudation are highly variable (Embleton and King, 1968: 251-258). In places the glaciers were more protective than erosive. Great valleys can be carved in highlands of relatively soft, sedimentary rocks. However, Ambrose (1964) and Flint (1971: 115-119) have emphasized the limited extent of Pleistocene glacial erosion (15 m or less) on the crystalline rocks of the low-relief, Canadian shield. The structural similarities between many of the Precambrian terrains of eastern Canada and the Precambrian terrains of the Lower Vaal Basin are striking, and net Paleozoic glacial erosion of eastern Griqualand West probably was equally limited. While it is difficult to separate the pre-glacial surface from the erosive effects of the Dwyka glaciers, it is reasonable to assume that its major shapes are the products of erosional processes and tectonic events prior to this glacial episode.

Structural Elements and Tectonic History

The shapes and locations of valleys and uplands on the Pre-Karoo Surface are directly related to the structure of the underlying Precambrian bedrock. These ancient structural elements are a prominent theme in the modern geomorphology, yet the structural geology and tectonic history of the Lower Vaal Basin have never been systematically investigated. All the Precambrian rocks exhibit extensive jointing and variable-displacement faulting.[15] However, the Dwyka shales and Karoo dolerites have much less prominent jointing. No post-Dwyka faults were discovered in this investigation.

---

[15]Most of the faulting in the Precambrian terrains involves displacements of only a few meters. The small fault in the road cut at the Barkly West bridge shows about 2 m vertical displacement; the fault planes are brecciated. Larger displacements occur along the lower Vaal. At Muirton there is a fault with a vertical displacement of about 90 m.

There are two, first-order, structural alignments in the Lower Vaal Basin -- the NNW-trending anticlines in the Ventersdorp rocks and the NNE-striking faults and master joints which parallel the Gaap Escarpment. The best-exposed anticline is the Spioenkop Anticline (see above), which runs from Hopetown past Douglas. It is paralleled 20 km to the SW by another anticline, here named the Pampoenpan Anticline, which barely intersects the modern landsurface. A similar anticlinal shape may be preserved between the Riet River gorge and the highlands along the Vaal north of Schmidtsdrif.[16] The axis of the Spioenkop Anticline undulates gently, and transverse fractures scar its flanks. The arching of this structure largely predates deposition of the Transvaal System, but facies changes in the Black Reef Series suggest movement may have been partly coincident with deposition of these rocks. Possibly much later the Gaap Escarpment was created. The Gaap Escarpment is not marked by one great fault, but by a myriad of lesser fractures that are only partly connected. Pre-Karoo diabase intrudes many of these fractures, while others were intruded by Karoo dolerites. This NNE-trend is also evidenced by the upland between the Vaal and Harts Rivers. The date of its elevation (and formation of the Kaap Valley) probably parallels the genesis of the escarpment.

A second-order structural alignment is examplified by a lineament that extends for over 90 km across the region. This lineament, here named the Dikbosch Fracture, occupies the Vaal channel immediately above and below Barkly West, extends WSW across Droogeveld, crosses the Vaal north of Schmidtsdrif and passes into

---

[16]Yet another anticline may plunge northwestwards from Kimberley. Clearly this fold pattern (with accessory faulting) explains the pattern of modern river junctions in du Toit's (1910) "Kaap Valley". The Vaal, Orange, and Brak Rivers all flow into the Kaap Valley from the southeast.

the Gaap Plateau (Fig. 3-12). In the Ventersdorp terrains along this route, the fracture usually (but not always) marks the course of a narrow, quartz porphyry dyke, which outcrops at the base of a discontinuous chain of shallow valleys. In Transvaal System terrains the fracture is marked by minor slip-strike displacement and diabase dykes. Since the quartz porphyry is probably of pre-Transvaal System age, the faults and intrusions along the escarpment indicate that this sheer zone was repeatedly mobilized.

Photo-lineaments are particularly prominent on all the pre-Karoo rocks (Fig. 3-13). They nearly always trend in the NNW, NNE, and WSW directions described above. However, determining their genesis from ground survey is difficult. Lineament valleys are usually filled with colluvial debris, Hutton Sands, or limestone tufas that bury bedrock contracts. Many lineaments mark faults on the flanks of the anticlines and masterjoints trending along the fold axes. Others mark faults, master joints and diabase dykes along the escarpment. (Many defy speculation.) In any case most lineaments are visible on photographs because they represent stripes of more densely jointed and fractured bedrocks. Subsequent differential erosion has excavated narrow valleys along the lineaments or, alternatively, they are marked by linear stripes of arboreal vegetation, which find deeper soils and greater moisture along their trace. One of the greatest lineaments, here named the Thorngrove Lineament, extends along the axis of the Vaal-Harts interfluve. It first appears in the region across the Vaal from Warrenton and then arcs southerly to pass north and west of Holpan.[17]

---

[17]Obviously the Thorngrove Lineament parallels the structural trend of the Gaap Escarpment, and is marked by many modern springs. Ancient spring deposits are locally well developed. It is worth speculating that the lower Harts valley may partly be a graben.

Fig. 3-12. Dikbosch Fracture (lower left to upper right) southwest of Barkly West; in the middle background is the gorge of the Vaal River above Gong-Gong (photo: K. W. Butzer).

Lineaments are rare and much more subtle on Karoo rocks. All the obvious lineaments appear to reflect dolerite and kimberlite dykes. Rhomboidal, master-joint networks occur on dolerite sills and on Dwyka rocks adjacent to dolerite sills and kimberlite fissures. Joint patterns in Dwyka rocks, which are not usually visible on air photos, seem to be closely related to stresses caused by dolerite and kimberlite intrusions. Pre-Karoo faults and master joints provided pathways for the dolerite and kimberlite intrusions. No doubt the joints in the Karoo rocks also reflect isostatic adjustments during the Cenozoic.

The history of structural stresses in the Lower Vaal Basin is summarized in Table 3-1. The length of this list, and the resultant, relatively minor deformations emphasize the persistent stability of regions sited near the center of ancient continental plates.

Despite the intrusions of the dolerite and kimberlite, there appears to have been no geomorphically significant, regional tectonic deformation of the Lower Vaal Basin since the deposition of the Dwyka Shales.[18] Bedding dips in the Dwyka shales are virtually horizontal, with only a slight inclination to the south. Wherever dipping or deformed beds are encountered, they are eminently compatible with bedding to be expected in their original deposition. In addition the Pre-Karoo photo-lineaments usually have not been translated upward through the Dwyka shales.[19] The implications are threefold. First, the modern Pre-Karoo Surface in the Lower Vaal Basin not only preserves the hills and valleys of a Paleozoic landscape, but exhibits them at essentially their original attitude. Second, late-Cenozoic erosional and depositional cycles cannot be explained by tectonic deformation within the region,[20] and third, as the ancient valleys have been re-exposed during the late Cenozoic, the modern streams adjusted to the structures of the Precambrian bedrock (Fig. 3-11), leading to re-establishment of pre-Karoo drainage lines.

---

[18]Interestingly there is a suggestion of continued uplift in post-Dwyka times along the Spioenkop Anticline north of Douglas. However, where the Vaal channelway crosses this anticline near Douglas, striated surfaces and tillites outcrop along the channel margins. It is axiomatic that fracture patterns, once established, are first selected to release later stresses. There are additional hints of minor deformation along the south flank of the Spioenkop Anticline northwest of Hopetown. Differential pressure release between Dwyka shales and adjacent dolerite dykes in the lower Harts River valley also seems to have introduced some low amplitude wrinkles.

[19]Photo-lineaments have been translated upward through Pleistocene tills on "aseismic" segments of the Canadian shield in Manitoba and Saskatchewan (see Mollard, 1957).

[20]This is contra Mayer (1973). The implications of apparent tectonic deformation outside the lower basin are considered in Chapters 4 and 10 below.

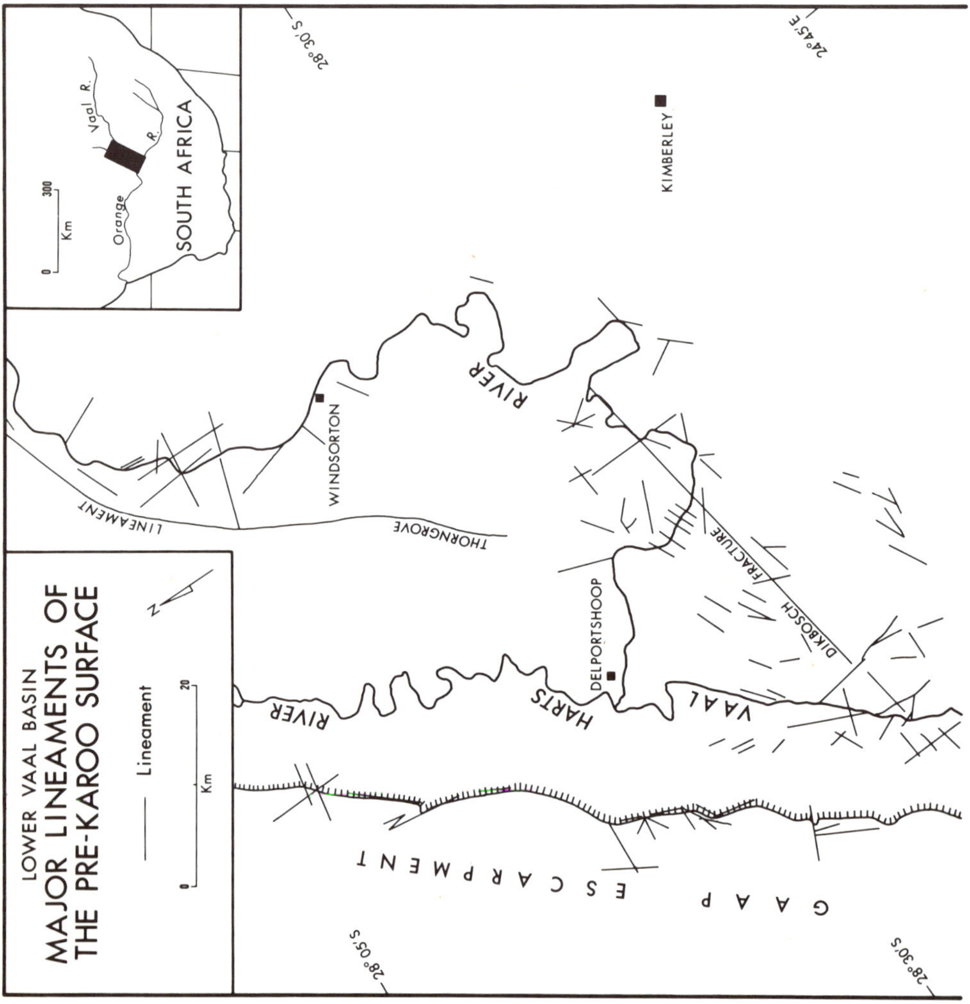
LOWER VAAL BASIN
MAJOR LINEAMENTS OF
THE PRE-KAROO SURFACE
Lineament
0
20
Km
N
0
300
Km
Vaal R.
Orange
R.
SOUTH AFRICA
28° 30' S
24° 45' E
KIMBERLEY
RIVER
WINDSORTON
THORNGROVE
LINEAMENT
DIKBOSCH
FRACTURE
DELPORTSHOOP
HARTS
RIVER
VAAL
G A A P E S C A R P M E N T
28° 05' S
28° 30' S

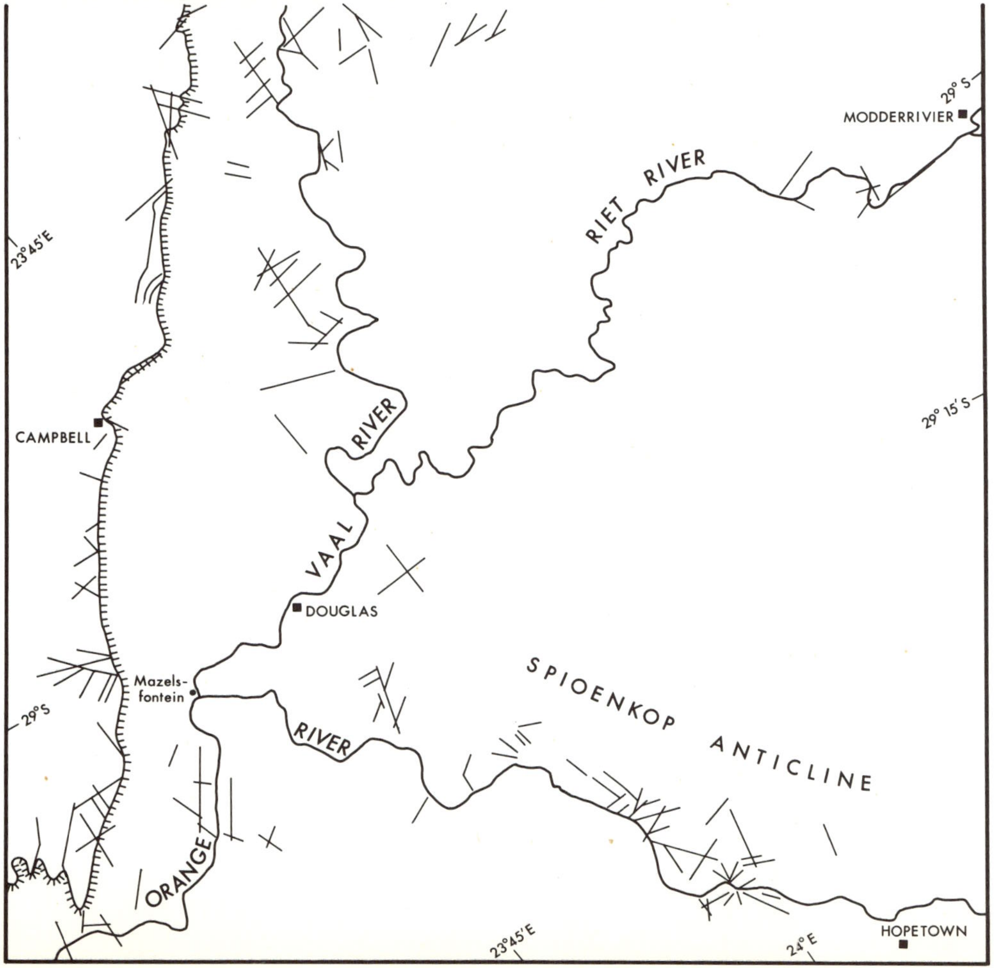

Fig. 3-13.
First-order lineaments in the Lower Vaal Basin.

TABLE 3-1

CHRONOLOGY OF STRUCTURAL STRESSES IN THE LOWER VAAL BASIN

12. Positive isostatic adjustments in response to Cenozoic erosion of the capping Karoo sedimentaries.
11. Kimberlite intrusions, multiple generations (?) (late Mesozoic)
10. Karoo dolerite intrusions, multiple generations (Triassic-Jurassic).
9. Negative isostatic adjustments in response to deposition of upper Karoo sedimentaries (Permian-Triasic).
8. Complex isostatic adjustments in response to the Dwyka glaciation (late Paleozoic).
7. Positive isostatic adjustments in response to extensive pre-Dwyka erosion.
6. NNE-trend faulting, particularly in Transvaal System rocks. Intrusion of diabase dykes, and (?) renewed folding in the Ventersdorp basement.
5. Transvaal System deposition and complex, coastal isostatic adjustments (ca. 2300 m.y. B.P.).
4. Spioenkop folding and accessary faulting, emplacement of pegmatites, quartz porphyries, and (?) diabase.
3. Ventersdorp volcanism and negative, coastal, isostatic adjustments (ca. 2400 m.y. B.P.).
2. Zoetlief volcanism and minor tectonism.
1. Multiple episodes of tectonism and volcanism in formation of the basal granites and schists.

Ridges and Valleys of the Pre-Karoo Surface

As the Pre-Karoo Surface was exhumed during the later Cenozoic, it has proved incredibly durable to continuing erosion. The exhumation has been accomplished by selective deflection of erosive energies to work on the softer Dwyka sediments. Striated surfaces and Dwyka fills frequently persist along both the planar and angular uplands, and glacial markers are present on the slopes leading into the river valleys. River channels in the Precambrian terrains regularly have tillites or striated surfaces along their margins (Fig. 3-11). Consequently the modern ridges and valleys in the areas of Precambrian bedrocks represent only slight modifications of the pre-Karoo topography and, in general, the Pre-Karoo Surface is visible wherever Precambrian rocks are exposed (Fig. 3-1). In this sense most of the slopes developed on Precambrian bedrocks are relict.

Maximum relief on the Pre-Karoo Surface occurs along the lower Orange, where the Gaap Escarpment stands over 400 m above the tillites at the base of the Orange River channel. In the uplands of Ventersdorp andesite, relief is typically much less and usually under 80 m. Most areas of the Pre-Karoo Surface undulate gently and local relief is commonly under 30 m. Planar segments of the Pre-Karoo Surface are locally well-developed.

The main pre-Karoo ridges developed in Ventersdorp andesites are the Vaal-Harts interfluve, the Spioenkop Anticline, and Nooitgedacht, Pniel, and Rooipoort uplands. Apart from the highest ridge crests the pre-Karoo slopes on the Ventersdorp volcanics are gently rounded with smooth, concavo-convex slope profiles. Low slope angles are common. Most drainage divides are broad and flat with a general southward dip. Modern erosional gullies are in near perfect adjustment with the joint system. These gullies are often less than one meter deep, and are themselves set in the center of local, flat-bottomed undulations where Dwyka sediments may also be preserved.

The extent of glacial modification of the uplands composed of Ventersdorp volcanics is difficult to estimate. Glaciers do not always leave behind diagnostic erosional forms and, as noted above, may be protective in many cases. Late-Cenozoic events have altered or buried many glacial features. The preservation of a thin, striated tillite on the Vaal-Harts interfluve argues not only for minor late Cenozoic changes, but for very little modification in the final glacial episode. Small, shallow pans are frequent on the low-relief ridge summits. These circular pans are typically located in tectono-erosional depressions which mark the junction of several fracture lines. It is also likely that some of the gullies on the upland surfaces mark the routes of subglacial channelways. In some areas irregularly-rounded, partly streamlined ridges rise above that flat upland surfaces. Where these ridges sit athwart the general glacial flow lines, the lee (usually south-facing) slopes are steeper and shorter. In addition, curious, hummocky slope shapes can be noted near some ridgecrests; with up to 10 m relief, they bear no relation to either pre-glacial or late-Cenozoic fluvial drainage lines. At present the slopes of these features are marked by fields of subangular, cobble- to boulder-grade andesite rubble. They appear to be due to some type of sub-glacial erosion, perhaps extensive plucking. Good examples are found on the highlands along both the east and west boundaries of Rooipoort (Fig. 3-14), and on the farms Doornhoek and Thorngrove to the west of Holpan. Among the highest parts of the Pre-Karoo Surface, these ridges endured not only extensive glacial erosion, but they were the first features to be exposed in the late Cenozoic and, as a consequence, have

Fig. 3-14. Boulder field along the andesite ridgecrest at the western margin of Rooipoort.

suffered the greatest amount of subsequent weathering and erosion. Perhaps these ridges were frost-sharpened before they were buried during Dwyka times.

At present most of the andesite uplands are veneered with a shallow regolith. Striated surfaces are preserved only on areas of nearly level terrain and on areas where the Dwyka sediments have only recently been removed. At many locations where glacially-striated and plastically-moulded features would be expected, weathering and hillslope erosion have produced slopes littered with subangular, cobble- to boulder-sized blocks of andesite. The surface wash and regolith in these areas is an unstructured, yellowish-red, loamy sand with occasional quartz pebbles. Many of the quartz pebbles are reworked amygdules. On low-angle slopes the matrix of the regolith may be calcareous. Corrosion of andesite surfaces has also occurred, e.g. the 1 to 2 m deep pits on the upland surface at Nooitgedacht (Chap. 4 below). On the whole however, enough exposures of Dwyka sediments and striated surfaces remain to demonstrate persuasively that only minor modification of the andesite uplands has occurred in the late Cenozoic.

Turning to the valleys in the andesite terrains, it is immediately apparent that not all pre-Karoo valleys are exposed. Important unexposed valleys are located west of Kimberley, along the Kareepan and Rooipan vloers (located west and south of Kimberley), and along the north side of the Spioenkop Anticline. These valleys cut across the anticlinal, structural grain. Major, exposed pre-Karoo valleys, cut in Ventersdorp rocks, include the Barkly West gorge, the Vaal River valley between Warrenton and Windsorton, sections of the Orange River valley from Hopetown to the Vaal confluence, and the Riet River gorge. Compared to the andesite uplands, these valleys have greater local relief and the valleyside slopes are steeper.[21] Profiles remain broadly concavo-convex while slopes are littered with subangular blocks of andesite and a veneer of sandy slopewash.

The frequent preservation of Dwyka tillites at the base of these valleys indicates that much of the shaping occurred before or during the late-Paleozoic glaciation. However, the valleys are often obliquely aligned with the large-scale glacier flow directions (particularly the Vaal valley between Nooitgedacht and Delportshoop), and glacial modification in these areas was more complex than simple deepening through selective scour. In some cases, shallow valleys from 1 to 3 km across, with associated striated "whaleback" ridges and valleyside fluting, can be traced entering or leaving the modern river valleys. Excellent examples are found along the

---

[21]For more slope data see Chapter 2.

Vaal at Nooitgedacht and south of Gong-Gong, and along the Orange near Ettrick. Where the pre-Karoo valleys were followed by the Dwyka glaciers, there are additional characteristics suggesting glacial erosion. These include sequences of basins and steps ("riegels") in the valley profiles. In the modern landscape the basins are still occupied by Dwyka sediments and the steps appear as striated rock bars just above the average channel level. Basins and steps are well developed on the Vaal between Windsorton and Riverton. The relief on the basins and steps suggest deepening of the presently exposed valleys during the glaciation averaged less than 20 m, while some of the unexposed valleys may be much deeper.[22]

The degree of late-Cenozoic modification of the pre-Karoo valleys in the andesite terrains is difficult to evaluate. Clearly a river that intermittantly carried cobble gravels had the power to erode its bed beyond the base of the pre-Karoo valleys. However, even in steep reaches with gravel terraces nearby, Dwyka sediments or surfaces can regularly be found either on or within a few meters of the bed of the modern channel. Despite this observation, localized and limited cutting below the bottoms of the pre-Karoo valleys has occurred. These include sites along the base of the main streams and along the base of tributary gullies. Such late-Cenozoic cutting has been limited to steep stream reaches and along lines of structural weakness. The depth of such localized cutting is probably less than 10 m.

The tabular uplands in the Transvaal metasedimentary terrains usually have veneers of Dwyka sediments around their bases and occasionally across their summits. Intervening valleys are narrow with flat bottoms and intimately adjusted to structure. Compared to the crystalline rocks of the Ventersdorp System, the sedimentaries of the Transvaal System are less durable and probably suffered

---

[22]In particular, the Spitskop borehole indicates the lower Harts valley was over-deepened by about 80 m.

greater modification during the Dwyka glaciation. A major glacier lobe moved down du Toit's "Kaap Valley" streamlining the topography in its path. The ridges west of the Vaal, such as the Mosesberg, are all drumloidally sculptured with striations and grooves along their summits. It is also likely that the smooth escarpment of the Gaap Plateau received its final trimming and shaping during glaciation (for an analogue with the Niagara cuesta, see Martin, 1916). At Paaiskloof and Witberg glaciers in hanging valleys (pre-glacial grabens) joined the main valley glacier (Fig. 3-15). Since the thickest exposures of tillite (greater than 300 m) are in the valley of the Orange below the Vaal confluence, it is likely that this area was considerably denuded during glaciation. However, striations and other small-scale features of glacial alteration are uncommon on the Transvaal System carbonate rocks. Fresh exposures often exhibit such markings, but modern weathering quickly erases the minor glacial scars.

Late-Cenozoic alteration of the pre-Karoo forms on terrains of Transvaal System rocks has been limited. Along the escarpment south of Ulco a few lateral streams have cut narrow chasms about 1 km back into the scarp front.[23] Near Schmidtsdrif lateral streams have probably deepened 2 to 5 m beyond their pre-Karoo levels. However, all deep valleys along the lower Orange still have thick accumulations of Dwyka sediments at their base or embanked against their slopes.

---

[23]The LANDSAT images of the escarpment suggest hemispherical scallops up to 5 km broad and perhaps 40 m deep have been cut in the crest of the escarpment during the later Cenozoic.

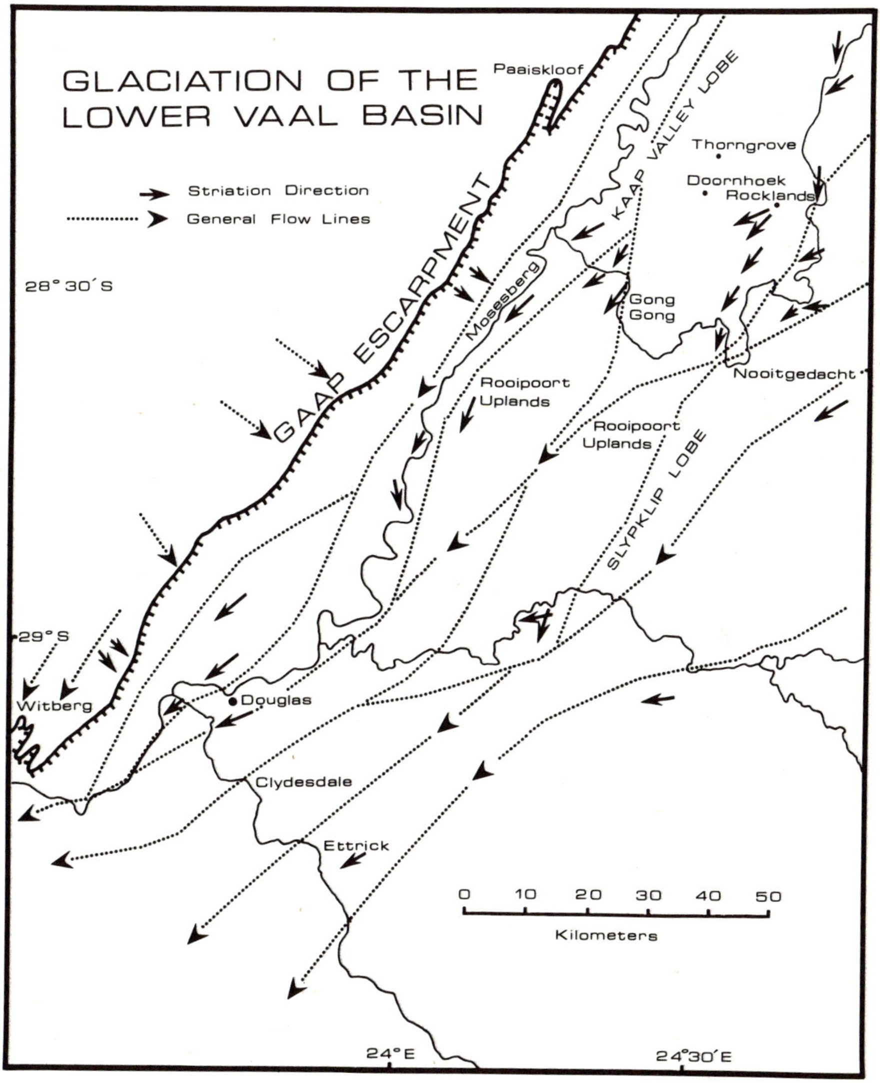

Fig. 3-15. Glaciation of the Lower Vaal Basin.

Pre-Karoo Drainage and the Dwyka Glaciation

The Permian glaciers which flowed across the Lower Vaal Basin rose in a highland area coincident with the modern Witwatersrand. A secondary source area was located west of the Gaap Plateau (Stratten, 1968). However, the southwesterly[24] lobes of the Witwatersrand center were of primary importance in the glaciation of the Lower Vaal Basin. For the Karoo Basin as a whole, the motions of the glaciers are fairly well known. Investigations of glacially-shaped valleys, tillite fabrics, and striations, as well as other small-scale gouges, have revealed the general directions of the glacial advance (see du Toit, 1953; Stratten, 1968). At the smaller scale provided by the Lower Vaal Basin, a greater variety of detailed movements can be recognized as the glaciers bent and shaped themselves to the preglacial topography. The following model is constructed from the shapes and alignment of only the erosional forms.

Although the region probably was continuously covered with ice, the main glacier lobes came from the northeast down the Kaap Valley and along the east side of the Vaal-Harts interfluve (here named the Slypklip Lobe). The Vaal-Harts interfluve acted as a medial ice shed, with ice flowing downslope on both sides, but predominantly into the Kaap Valley. Near Nooitgedacht and Barkly West, arms of the Slipklip Lobe rode down into the bends of the pre-Karoo river valley only to exit on the opposite side. The Slypklip Lobe continued southwesterly and along the depression of the Rooipan vloer, and linked with the Kaap Valley Lobe north of Douglas. Near Gong-Gong an arm of the Kaap Valley Lobe flowed into the pre-Karoo river valley, and, relative to it, flowed upstream

---

[24]Direction references are obviously justified to modern polar positions.

(still southerly) for a distance of several kilometers before climbing out of the valley to join the Slypklip Lobe. The Spioenkop Anticline did little to deflect the glaciers and was overrun along its whole length. On its lee (southerly) side the glaciers scoured u-shaped valleys leading into the modern Orange River. At the termination of the Dwyka sedimentation all the pre-Karoo valleys had been filled beyond their rims.

The detailed ice movements are much more complex. The directions revealed by striations and gouges are more often a reflection of local topography than regional ice movements. It appears the glaciers were more deformed by the pre-Karoo terrain than vice versa.[25] The directions recorded by striations nearly always have a southerly component, but the other component can be east or west. At large striated outcrops, the striation directions often vary through an arc of more than 80°. The general directions indicated by Dwyka erosional features are shown on Figure 3-15.[26]

It must be assumed that the glaciers followed pre-Karoo river valleys. These river channel positions can be modeled as follows (Fig. 3-16). Channels of the pre-Karoo rivers are marked in part by the modern Vaal, Harts, Riet, and Orange Rivers. A major stream, here named the "pre-Dwyka Vaal", flowed from the northeast to the modern Vaal channel at Windsorton (noted by du Toit, 1910). The modern Vaal above Windsorton only marks a tributary channel of this

---

[25]Perhaps an indication of subaqueous erosion.

[26]The presence of banded ironstones and jaspers in tillites along the Riet River gorge indicates that the glacier lobes trailing southeastward across the Langeberg and the Gaap Plateau areas were also important in the glaciation of the region. However, there appear to be no coincident erosional forms excepting equivocal "stone stripes" visible in air photos at several sites along the escarpment crest and the crest of the Mosesberg.

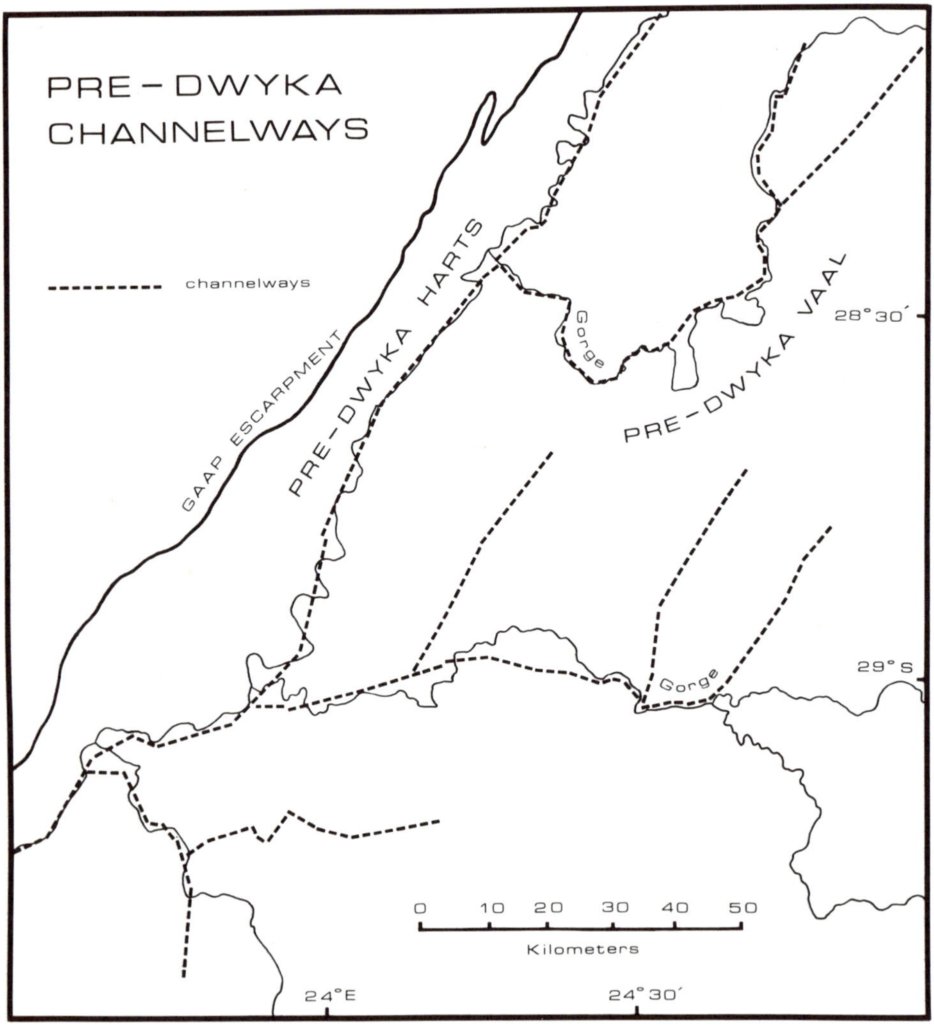

Fig. 3-16. Pre-Dwyka drainage lines.

high-order stream. The pre-Dwyka Vaal continued south, generally following the modern Vaal through the Barkly West gorge and entered the major river in the Kaap Valley south of Delportshoop (contra du Toit, 1910). The Kaap Valley stream, here named the "pre-Dwyka Harts", flowed east of the modern Harts above Delportshoop, but generally coincided with the modern Vaal south of Delportshoop. Further downstream the pre-Dwyka Harts received another left-bank tributary which paralleled the modern Riet. This stream had a complex network with one branch reaching up the Rooipan vloer and another reaching up the Riet River gorge. South of Douglas the pre-Dwyka Harts received yet another left-bank tributary. This stream paralleled the modern Orange as far as Torquay. The channel of the Orange River below the Vaal confluence is identical to that of the pre-Dwyka Harts as it passed southward out of the region.

Whether the channel of the Orange River between Hopetown and Torquay marks a pre-Dwyka river is an unresolved question. A stream running somewhere in the synclinal valley between the Spioenkop and Pampoenpan anticlines is to be anticipated. In general, the Orange channel in this reach has Dwyka sediments above its left bank, with Precambrian rocks at low levels on the left bank, on the channel bottom, and above the right bank. A channelway may be buried to the south of the present stream. Where the Orange is deeply incised in Precambrian rocks, such as upstream of Torquay, this cutting may represent scour during glaciation and/or selective late-Cenozoic erosion along lines of structural weakness.

Planar Features of the Pre-Karoo Surface

The exhumation of the pre-Karoo landscape has revealed several ancient planar features, which now form low-relief surfaces in the modern landscape. Notable among these are parts of the Gaap Plateau and the suite of platforms along the course of the pre-Dwyka Vaal River (Fig. 3-17).

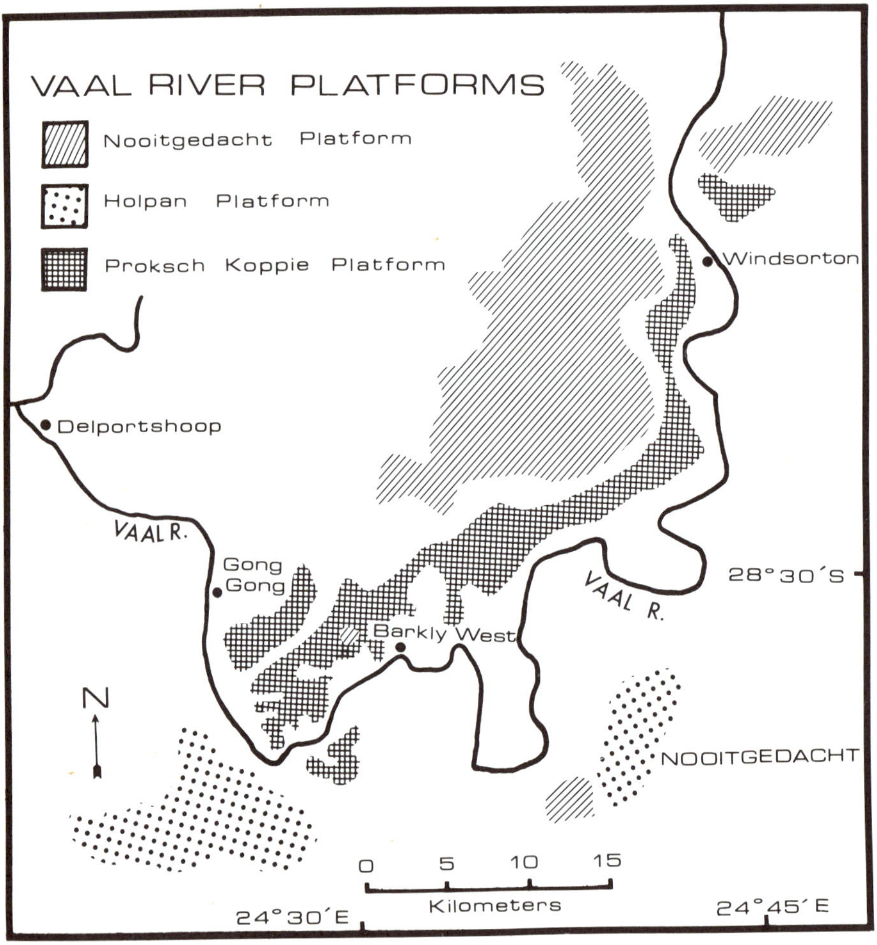

Fig. 3-17. Pre-Karoo platforms along the lower Vaal River.

At Paaiskloof (north of Ulco) and at Witberg (on the south edge of the Gaap Plateau) Dwyka shales bank up and over the Gaap Escarpment, clearly demonstrating that the escarpment is a pre-Karoo relict.[27] At these sites the planar escarpment crest also emerges from beneath Dwyka sediments, and these surfaces are obvious pre-Karoo relicts. But it is unclear to what extent the much larger plateau surface should be viewed as a pre-Karoo feature.[28] With a width of about 60 km, the Gaap Plateau starts near the 1200 m contour along the escarpment, and rises gently westward to about the 1500 m level (a regional slope of about 0.5%). The widely-spaced surface drainage is generally limited to shallow locally discontinuous channels that run southeastwards, against the trend of the modern Vaal system. Inclusions of Karoo rocks in the kimberlite pipes on the plateau indicate the former presence of a Karoo sedimentary cover near the level of the modern plateau surface. However, wide pre-Karoo plains remain to be demonstrated. Bedrock outcrops on the surface are usually obscured by veneers of limestone tufa and calcrete. As a karstic terrain composed of westerly-dipping dolomitic limestones, it is difficult to envisage any set of planation processes which could have produced the minimal-relief plateau surface. It is equally difficult to postulate a means of removing the Karoo System sediments once covering this region, while still leaving the surface only scarred by shallow polje

---

[27]Dwyka sediments are also well developed in some areas along the escarpment base, particularly northwest and southwest of Douglas.

[28]The south margin of the escarpment also corresponds to a post-Karoo erosional bevel. In this area the escarpment is developed exclusively in Dwyka shales and tillites and capped by an early Cenozoic (?), limestone tufa. This tufa dips southward at about 1° and is 6 m to 10 m thick. On morpho-stratigraphic grounds it is the oldest element in the surficial geology of the region.

drainage networks. The origin of the Gaap Plateau and the nature of late-Cenozoic denudation on this surface remain enigmatic problems in the geologic history of the region.

Planar surfaces, often sporadically coated with Cenozoic fluvial and lag-concentrated gravels are well preserved along the Vaal between Windsorton and Gong-Gong and along the Riet River gorge. However, these surfaces, which are developed in rocks of the Ventersdorp System, also frequently preserve glacial-age erosional markers and local fills of Dwyka sediments. The abundance of relict glacial features, either on the surface or around their margins, indicates that the ultimate origin of these surfaces must be sought during or before the Dwyka glaciations. These surfaces are relatively flat, but relief is available around small, streamlined, koppies, glacially-scoured depressions, and shallow, late-Cenozoic valleys. The surfaces generally trend down the axes of the pre-Karoo channelways, and their boundaries are marked by gentle, poorly defined convexities.

The Paleozoic plains along the Vaal are named the Nooitgedacht, Holpan, and Proksch Koppie Platforms.[29] The Nooitgedacht Platform is found at two localities, both of which are beyond the left bank of the modern Vaal. One locality is the highlands on the farm Nooitgedacht, where it is covered by late-Cenozoic fluvial gravels, and the other, which apparently has no gravels, is located south of Gong-Gong at the junction of the farms Pniel and Rooipoort. At Nooitgedacht the platform is at an average elevation of 1180 m (85 m above the modern Vaal), and at a general elevation of 1165 m on Rooipoort, a drop of 15 m in 22 km or approximately 0.7 m/km.

The extensive Holpan Platform is subcontinuous along the crest of the modern Vaal-Harts interfluve. In addition, two more parts are found east of the modern Vaal at Nooitgedacht and north of

---

[29]These features were first named by Butzer, Helgren, *et al.*, (1973).

Windsorton. At its northern extremity this platform reaches an elevation of about 1170 m (65 m above the modern Vaal) and declines to 1145 m above Barkly West, a drop of about 25 m in 25 km or approximately 1 m/km. Limited relief on the Holpan Platform includes several small koppies which rise up to 25 m above the general surface level.

Downslope from the Holpan Platform is the more complex Proksch Koppie Platform, which extends along the Vaal's right bank from Windsorton to below Barkly West. Outliers of this surface to the east of the Vaal are found north of Windsorton and west of Riverton. The surface extends from an elevation of 1140 m near Windsorton, where it is 30 m above the modern Vaal, to an elevation of 1120 m south of Barkly West, where it is 70 m above the modern Vaal, a drop of about 20 m in 35 km (0.6 m/km). Above Gong-Gong there is a 15 m lower segment of this surface; its upper margin is a lineament which may mark a Paleozoic fault or the former site of igneous dyke that inhibited Paleozoic river incision.

While the origin of these surfaces probably is planation to local base levels, the origin of the Nooitgedacht and Holpan Platforms as topographic highs of the Pre-Karoo Surface must in part reflect even older tectonic events. The nearly equal height of the Gaap escarpment and the Holpan Platform cannot be ascribed to random processes. Yet their positions, and particularly the position of the Proksch Koppie Platform on the flanks of the valley of the pre-Dwyka Vaal River, suggest that they indicate cyclic episodes of downcutting in the Paleozoic. It is unknown whether this planation preceded or was part of the Dwyka glaciation. Contrary to the above arguments, Cooke (1947, following Wagner, 1914) suggested that the course of the Vaal River below Barkly West reflects late-Cenozoic capture of the Vaal by a tributary of the Harts. It is probable that the channel of the Vaal in the Barkly West gorge does reflect river capture, but such capture must have occurred during late-Paleozoic times. Probably prior to this capture, the pre-Dwyka Vaal flowed down the valley now marked by the modern Rooipan vloer to join the pre-Dwyka Harts north of Douglas.

Dwyka tillites and shales fill both the upper and lower ends of the Riet River gorge, and glacially moulded surfaces repeatedly trail in and out of this narrow valley.[30] The modern Riet River has exhumed an impressive, incised valley network. Slightly concave surfaces, which preserve glacial markers, connect the crest of the gorge with the more distant, shale plains. These pre-Karoo paleo-surfaces were probably eroded prior to incision of the gorge (Fig. 7-4). Developed at 1120 m to 1100 m these surfaces descend less steeply than the gorge. Relative elevations increase down-valley from +25 to 45 m.

Implications

This outline of the geologic history of the Lower Vaal Basin provides a background for description and interpretation of the much younger Vaal River terraces and alluvial deposits. There are two critical implications. First, a detailed examination of the pre-Karoo landscape shows that the steep reaches of the Vaal valley, particularly those between Warrenton and Windsorton and between Barkly West and Delportshoop, are exhumed pre-Karoo valleys. These valleys were filled with relatively soft, Dwyka sediments, which were selectively removed during Cenozoic exhumation. These sediments could not have formed the lithologic "barriers" to erosion thar were formative in the interpretations of regional Cenozoic denudation proposed by du Toit (1907), Söhnge and Visser (1937),

---

[30]The gorge is about 24 km long. The Riet River drops 60 m in this distance, forming the steepest reach of the Lower Vaal Basin.

Cooke (1947), as well as Partridge and Brink (1967). And secondly, there are no indications of significant tectonic deformation within the Lower Vaal Basin during the Cenozoic.

CHAPTER 4

THE OLDER GRAVELS OF THE VAAL RIVER

Around the green gravel the
grass grows green,
Wash them with milk, and clothe
them with silk,
And write their names with pen
and ink.

--after Mother Goose

Introduction

During the Cenozoic the rivers of the Lower Vaal Basin cut their way through Karoo sediments exposing parts of the Pre-Karoo Surface. The first Cenozoic events are recorded by fluvial and colluvial, rudaceous deposits collectively known as the "Older Gravels".[1] These Older Gravels are best developed along the Vaal River between Windsorton and Barkly West, but comparable deposits and terraces are found along other reaches of the Vaal as well as along the Riet, Harts, and Orange rivers. Small-scale mining activities over the last century have removed all but a few of the original Older Gravels deposits, and most of the Older Gravels

---

[1]The terms "Older Gravels" and "Younger Gravels" are here used as informal lithostratigraphic units to denote a range of features that may or may not include gravel. As a noun or adjective, e.g. "Older Gravels platform", the term is considered a plural form for linguistic facility, without implying multiple subunits.

terrains now are little more than clusters of low mounds of carefully sorted gravels. The rare, freshly mined exposures of shallow Older Gravels are remarkably similar to one another. Typically, such sites reveal less than 2 m of uncemented, rubefied gravels that fill irregular, rounded or conical solution cavities in densely calcreted gravels; the diggers remove the loose rubble, leaving sorted heaps of waste gravels around exposed cavities, which are up to 6 m wide and 2 m deep.

The term "Older Gravels" was first specifically used in the pioneering study of Söhnge and Visser (1937), who applied the label to all gravels altimetrically higher than a complex of gravel terraces frequently found adjoining the modern Vaal River; the lower gravel terraces were designated the "Younger Gravels". These two classes of diamondiferous deposits were also distinguished by the frequent recovery of Acheulian artifacts from many Younger Gravels outcrops.

The Older Gravels, although not yet named, were well-known long before the survey of Söhnge and Visser. The first rush of diamond diggers in the late 1860's had already explored many exposures before the visit of Stow in 1872, and by 1914 Wagner (pp. 263-270) distinguished between "high level" and "low level" gravels. Although many Older Gravels deposits, such as those at "Klipdam", "Pniel", and the "Droogeveld Channel", became famous for their diamond production, the matter of their origin and systematic interrelationships was largely neglected. Many exposures were explicitly recognized as former channel deposits, but the first and only attempt to correlate these gravels with a denudational sequence was made by Williams (1932: 542-546), an account dealing mainly with deposits outside the Lower Vaal Basin.[2]

---

[2]Some important Older Gravels deposits are located outside the Lower Vaal Basin. Most significant among these are the "London Run" between Schweizer-Reneke and Bloemhof, the gravels in the Lichtenberg

The primary area for examination of both the Older Gravels and the Younger Gravels has been near Windsorton (Figs. 4-1, 4-2). There the Older Gravels spread laterally for over 10 km. Near Windsorton, Cooke (1947) and van Riet Lowe (in Breuil, et. al., 1948) recognized a simple sequence of post-depositional alteration in the Older Gravels. These conglomerates were divided into two segments -- the capping "Red Older Gravels" or "Potato Gravels" and the calcreted "Basal Older Gravels". Cooke (1947) argued correctly that the uncemented and rubefied "Red Older Gravels" were a weathered and colluvially reworked residual of the calcreted "Basal Older Gravels". However, the "Red Older Gravels", being a concentrate, contained a greater proportion of diamonds and, as a result, have been almost entirely mined away. The next systematic examination of the deposits near Windsorton was by Partridge and Brink (1967). They amplified on Cooke's classification by recognizing that the "Red Older Gravels" had been subject to substantial local redistribution, and re-labeled the "Red Older Gravels" as "Derived Gravels". They also re-labeled the "Basal Older Gravels" as "Primary Alluvial Gravels". Significantly, Partridge and Brink were the first to explicitly identify "terraces" associated with the Older Gravels. They identified three "terraces" at elevations of 70 feet, 100 feet, and 200 feet above the modern Vaal. Butzer, Helgren, Fock, and Stuckenrath (1973) expanded the "terrace" sequence for the region south to Barkly West and identified three

---

district, and the gravels near Vereeniging. Other, minor outcrops of Older Gravels are preserved in small discontinuous patches across much of the Vaal-Limpopo watershed. For locations and descriptions see Wagner (1914), Williams (1932), Söhnge and Visser (1937) and du Toit (1951). These deposits show that the Vaal system had both a different drainage network and a larger basin during much of Older Gravel times. However, none of the Older Gravels of the upper basin appear to be as young as the Wedburg Terrace, and their comparability with the other Older Gravels in the lower basin is obscure.

Fig. 4-1. Older Gravels terrain near Holpan I; a calcreted, primary deposit is at left.

Fig. 4-2. Diamond diggers excavation near Holpan I; bedrock at left is a roche moutonnèe of andesite. Tunnel at center is 1.6 m high.

"fluvial platforms" associated with Older Gravel deposits. These are the Proksch Koppie Platform -- which is equivalent to the 100 foot terrace, the Holpan Platform -- which is equivalent to the 200 foot terrace, and the Nooitgedacht Platform -- which is located east of Barkly West and not present near Windsorton.

This investigation has shown these platforms (as well as the 100 foot and 200 foot terraces at Windsorton) must be regarded primarily as pre-Karoo features remodelled by rivers during late-Cenozoic exposure of the Pre-Karoo Surface (see Chap. 3 above). It is also apparent from this investigation that the "70 foot terrace" at Windsorton has wide parallels throughout the Lower Vaal Basin; it is here designated as the Wedburg Terrace. Finally, this investigation serves to show that deposition of the primary Older Gravels as well as their redistribution were more systematic and complex than perceived by Partridge and Brink (1967).

In this discussion the spatial context and sedimentary characteristics of each set of Older Gravel deposits will be identified and described. The incomplete historical record revealed by the Older Gravels is outlined in the concluding evaluation.

Nooitgedacht Gravels

Above the Wedburg Terrace in the Windsorton-Barkly West sector are the Nooitgedacht, Holpan, and Proksch Koppie Older Gravels (Fig. 4-3). At an elevation of 1175 m, the Nooitgedacht gravels are the highest in the Lower Vaal Basin. These gravels are spread across a pre-Karoo platform of Ventersdorp andesite, 85 m above the glacial pavements along the modern Vaal. Resting on the highest surface and exhibiting the most intensive, post-depositional weathering, the Nooitgedacht gravels can be regarded as the oldest rudaceous deposit in the Lower Vaal Basin.

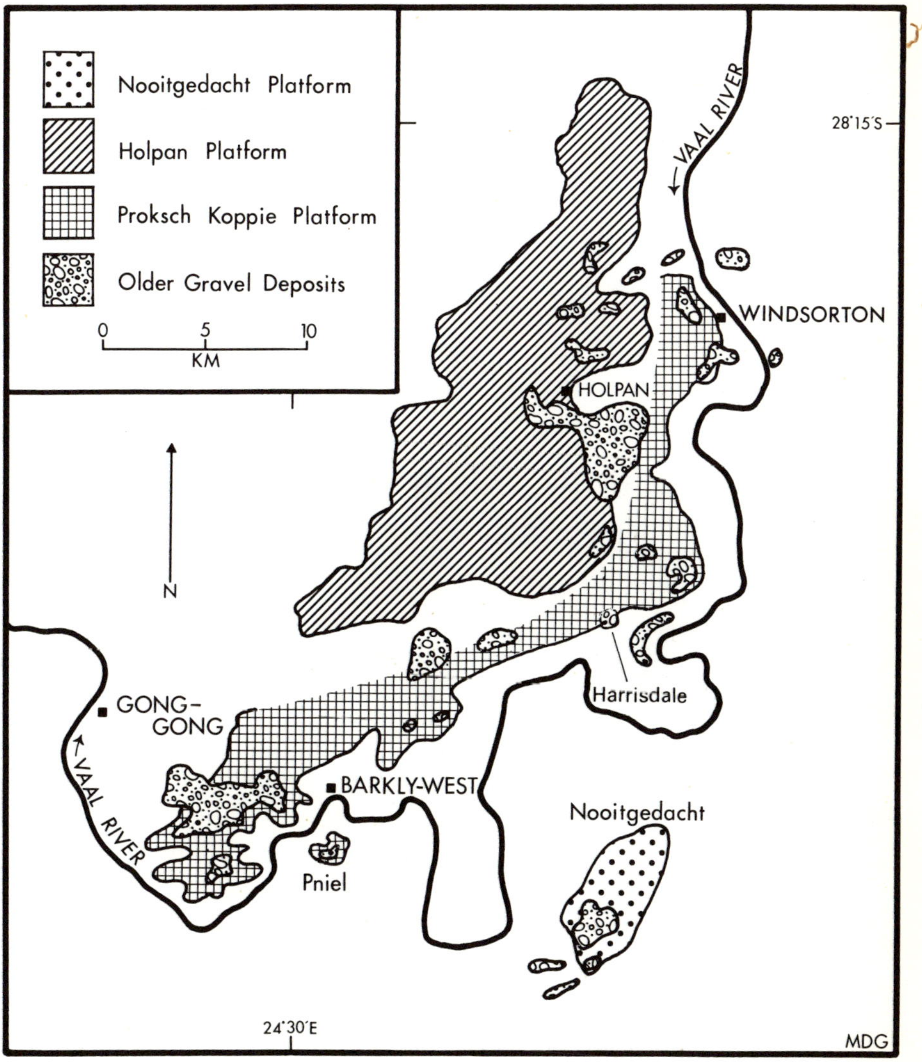

Fig. 4-3. High-level Older Gravels between Windsorton and Gong-Gong (only the largest deposits are mapped).

Although the Nooitgedacht gravels extend over an area of 5 km, only a few undisturbed deposits remain. The typical gravel exposure is only 10 to 20 cm thick, but the present diamond diggers report that some occurrences, now mined out, were up to 3 m thick. Gravel dumps near the sites of these thicker sections were examined and there is no reason to suspect that these deeper gravels were any more complex than the following type section (Figs. 4-4, 4-5; for sedimentary definitions and further data see Appendix A).

NOOITGEDACHT

Unit 1. (top) "Sands", 0.5-1.0 m, very well sorted, dark red (2.5 YR) fine sand, which is apedal without distinct horizonation (no. 001).

-- diffuse, irregular, unconformable contact --

Unit 2. "Nooitgedacht Gravels", 0.1 to 0.3 m, uncemented, granule to coarse-pebble conglomerate. The gravels are subrounded to rounded resistent lithologies, i.e. quartz, quartzites, agates, fossil wood, and gneiss and schist "bantams" (Wagner, 1914: 251 for a definition), mixed with angular to sub-angular pebbles and cobbles of locally derived andesite. The matrix is dark red (2.5 YR), well sorted, fine to medium sand (no. 002).

-- abrupt, irregular unconformable contact --

Unit 3. "Bedrock", Ventersdorp andesite, with an unevenly pitted surface and relief of up to 1.5 m, due to differential weathering along joints.

Unit 1 sedimentary characteristics are analogous to the Hutton Sands (see Chap. 2 above). In addition, nearby surfaces (particularly north of the railway line) exhibit the subdued eolian slopeforms typical of Hutton Sands terrains. However, the color is a deeper red than usual, suggesting additional sesquioxides derived from local weathering of the andesite bedrock.

Fig. 4-4. Diamond-diggers excavation at Nooitgedacht; corroded (cleaned) andesite bedrock in foreground, Hutton Sands veneer at left. In the background is a turn-of-the-century, diamond-sorting rig still in use in 1973.

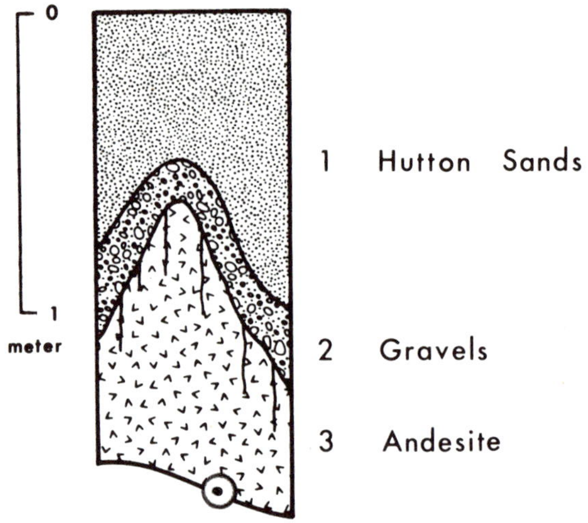

Fig. 4-5. Generalized sedimentary profile at Nooitgedacht.

The gravels of Unit 2 are not primary. The dominance of rock-types most resistant to chemical weathering as well as gravel in-fillings of the irregular andesite surface indicate colluvial redistribution and intervening weathering are responsible for their present condition and position. The latest redistribution was partly coincident with deposition of the Hutton Sands, as suggested by the similarity between the gravel matrix and the capping sands. However, the frequency and distance of previous redistribution cannot be determined.

The value of the Nooitgedacht gravels for paleo-environmental inference is minimal. They contain no record of the time, place, or character of their initial, presumably alluvial aggradation. This deposit is simply a weathered lag derived from an earlier alluvial deposit, aggraded before exposure of the pre-Karoo plat-form. Whether this former alluvium was deposited by an ancestral

Vaal River of by a tributary is unclear, since such gravels could have been derived "ready made" from a tillite, and also there is no way to reconstruct flow directions from existing sedimentary structures. Even the presence of fossil wood in the gravels presumably from the Ecca series, requires no transport from outside the region. Sub-horizontal Ecca sedimentaries are present at absolute elevations of only 1250 meters, immediately east of the Lower Vaal Basin, and at one time these Ecca beds extended across all of the region. No fossils or artifacts have been recovered from the Nooitgedacht gravels.

Also present near Nooitgedacht are thin, slope-wash gravels found on the southwest side of the Nooitgedacht Platform. Except for local matrix calcretion these gravels are lithologically analogous to those on top of the platform, and presumably have a similar origin, except that they rest on a pre-Karoo slope; they have been more extensively mobilized during successive intervals of hillslope erosion.

Holpan Gravels

The Holpan Platform, the next-lower surface on which Older Gravels are found, is located along the crest of the Vaal-Harts interfluve, extending from west of Windsorton to north of Barkly West (Fig. 4-3). Surface area exceeds 160 km^2 and at Holpan this feature is 60 m above the Vaal. Thin, colluvially re-worked, surficial gravels (i.e. "Red Older Gravels") are common over much of this surface, but deposits greater than one meter thick are found only rarely -- such as near Holpan and in local gullies. The surficial gravels, as well as any uncemented deeper deposits, have been excavated and resorted by diamond diggers so that only a few fragmentary deposits of the surficial gravels still exist. However, the bottom part of thicker gravel exposures typically is cemented (i.e. "Basal Older Gravels") and has been difficult to remove by diggers (Figs. 4-1, 4-2). The primary area of thick gravels is along the base of a gentle swale just south of Holpan, where the following exposure, named Holpan I, records the maximum variability in this area (Fig. 4-6).[3] Gravel morphometry data appear in Appendix B.

HOLPAN I

Unit 1. (top) "Derived Gravels 1A", a rubefied, uncemented granule to coarse-cobble conglomerate analogous to the "Red Older Gravels". Thickness averages 1.2 m, but the base is characterized by cone-shaped rubefication zones, locally penetrating an additional 0.9 m downward. The dark red (2.5 YR) matrix is unstructured, uncemented, and well sorted silty sand (no. 146).

-- abrupt, irregular weathering boundary --

Unit 2. "Derived Gravels 1B," 1.1 - 1.6 m, strongly-cemented, granule to coarse-cobble conglomerate with local sand lenses analogous to the "Basal Older Gravels". The crudely-bedded, subrounded gravels dip at 9° and strike 80° to 100° east of north (toward the modern Vaal) and clearly show it to be a fill unit. The typical matrix is pink (5 YR 8/6), very poorly sorted, strongly-cemented, silty sand, (no. 147) with secondary mesocrystalline calcite growths. Local sand lenses up to 20 cm thick are pale yellow (2.5 YR 8/3), very well sorted, medium sands (no. 148) and yellowish red (5 YR) fine sands.

-- abrupt, smooth unconformity --

---

[3]It is possible that non-rudaceous deposits comparable to Holpan I remain unexposed near Holpan. Several local farmers report that bore-holes pass through 2 to 10 m of "calcrete" before andesite bedrock is reached. Such thick "calcretes" in the study area tend to be cemented, fine-grained, fluvial, colluvial, or lacustrine deposits.

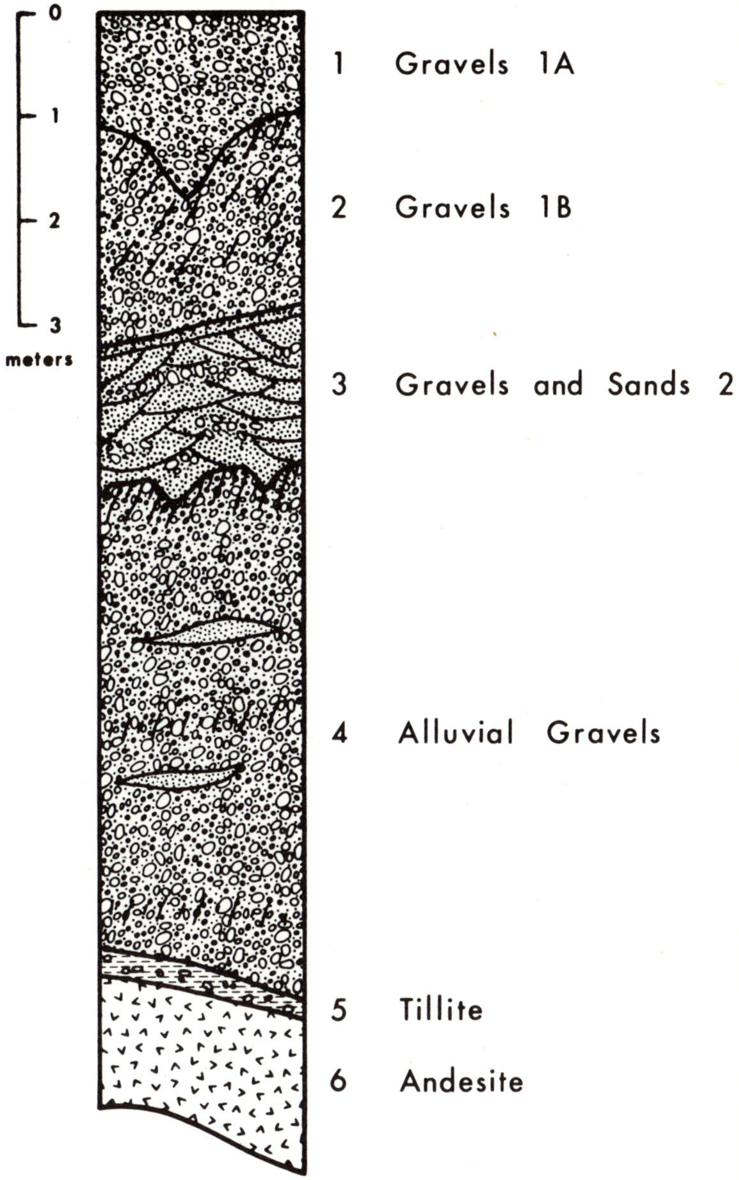

Fig. 4-6. Generalized sedimentary profile at Holpan I.

Unit 3. "Derived Gravels and Sands 2", 0.6 - 1.6 m, generally uncemented sand with tabular and trough cross-bedding and dispersed, fine pebble to coarse cobble-grade, subangular to rounded gravels. The top of the unit is a 5 to 8 cm, horizontal layer of white (2.5 YR 10/2), consolidated medium sand (no. 149). The typical cross-bedded sand of this unit is moderately sorted, very pale brown (10 YR 8/3), medium to fine sand (nos. 150 and 151). At the base is a narrow band of yellowish red (5 YR) fine sand (no. 152) which fills solution cavities up to 30 cm deep in the calcreted surface of unit 4.

-- irregular, abrupt uncomformable contact dipping at 6°, striking 80° to 100° east of north (toward the modern Vaal) --

Unit 4. "Alluvial Gravels", 4 - 6 m, variably cemented, granule to boulder-grade conglomerate with crude horizontal bedding and thin, local sand lenses. The matrix (nos. 153, 154) is poorly sorted, medium sand to moderately sorted, coarse sand (Fig. 4-8). One local sand lense (sample 155) is very poorly sorted medium and fine sand. Colors range from white (2.5 Y 8/1, N 10/0) to pale yellow (2.5 YR 8/4) to reddish yellow and pink (5-7.5 YR 6-7/4-6). Alluvially sorted, medium-coarse sands irregularly occur along the gravel base.

-- irregular, abrupt unconformable contact --

Unit 5. "Bedrock", 30 - 50 cm, dark, brown, striated, Dwyka tillite over plastically deformed and striated "roche moutonées" of Ventersdorp andesite.

The Holpan I exposure shows much greater variability than has previously been recognized in the Older Gravels. At the base of the sequence are striated tillite and andesite. Above the bedrock is a layer of gravels ("Alluvial Gravels"). Uncomformably above these basal gravels is the inhomogeneous and complexly-bedded sands and gravels, which comprise the first, major fill ("Derived Gravels and Sands 2"). Unconformably above this fill deposit is the second,

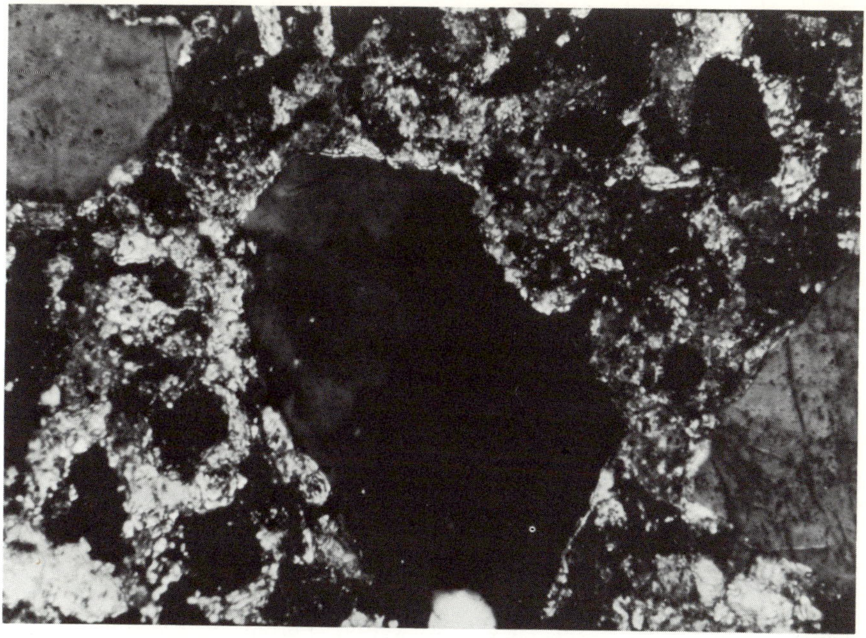

Fig. 4-7. Photomicrograph of sample no. 153, the calcreted gravel matrix at the surface of the "Alluvial Gravels" at Holpan I. Quartz grains are set in a matrix of very fine-grained, rather "dirty", calcite. Frame width is 0.65 mm.

major fill of derived gravels ("Derived Gravels 1A" and "1B"). Thus there are two major fills above what appear to be authentic, basal "Alluvial Gravels".

The upper part of the Derived Gravels has variable characteristics. At Holpan I, Unit 1 is an intact, decalcified example of Unit 2. However, in nearby exposures the gravels of Unit 1 have been colluvially reworked and resorted to emphasize both sand and gravel components. This debris fills solution cavities in the calcreted surface of Unit 2. Such reworking of the "Red Older Gravels" must be explicitly assumed to explain artifact occurrences both in and on this unit as reported by van Riet Lowe (1937, 1952), Breuil (1948), and Mason (1967).

The changes in local geomorphic environment recorded in the Holpan I exposure are impressive, even though they lack temporal references. The chronology begins with erosion of the Karoo sedimentaries once above the Holpan Platform. The preservation of striations on both tillite and andesite indicates the ineffectiveness of post-glacial erosional processes, even though a major river must have flowed over at least parts of the Holpan Platform to deposit the "Alluvial Gravels". These suggest a mid-channel, bedload deposit, now preserved in a glacially carved andesite hollow. Du Toit (1907) first pointed out that the modern Harts River valley could not have existed at this time, and that the Gaap Escarpment had only begun to be exhumed. Subsequently the major river gradually slid eastward, down the pre-Karoo slope, and no longer directly affected the sediments upon it. Erosional processes, probably in the channel of an ephemeral(?), tributary stream, then removed part of the "Alluvial Gravels" to leave a smooth erosional bevel on its surface. The "Alluvial Gravels" were later cemented with calcareous precipitates (i.e. calcreted). During a subsequent period, solution cavities formed in the calcreted surface of the "Alluvial Gravels". The residual surface sediments created at this time were yellowish red, fine sands, which now fill these cavities. Subsequently, fluvial processes were reactivated at Holpan I with net aggradation of "Derived Gravels and Sands 2", the composition and bedding of which suggest a torrential ephemeral stream, whose headwaters tapped sands and gravels comparable to the "Alluvial Gravels". A brief depositional hiatus followed, with limited erosion and weak calcretion of the top of "Derived Gravels and Sands 2". The end of this hiatus was marked by rapid aggradation of the "Derived Gravels 1A and 1B". At this time Karoo sediments may have been much more common than at present on the Holpan Platform, possibly providing some local relief to help generate the stream velocities implied by the deposition of this last fill. The "Derived Gravels" were then calcreted, and still later partly decalcified and rubefied through selective weathering of the ferro-magnesian minerals in the igneous clasts. Coincident, in part, with this rubefaction was local colluvial remobilization of the gravels near the surface.

An interesting characteristic of the Holpan I exposure is the repeated availability of fine, yellowish red sands with thin lenses preserved in all three depositional units. These sands are similar to the Hutton Sands, whose genesis is discussed in Chapter 10 below.

The cut-and-fill events recorded at Holpan I indicate radical departures from contemporary fluvial events on the Holpan Platform. Gravels comparable to Unit 2 are not mobilized by extreme modern, hydrological events, and only minor redistribution of the Unit 1 surface is now possible. Modern drainage on the platform is carried by an unintegrated system of shallow channelways, seldom more than 2 m wide. Since there has been no significant tectonic activity in the Lower Vaal Basin during the late Cenozoic, these cut-and-fill events must be attributed to repeated and significant changes of hydrological regime. In the context of low topographic diversity on the Holpan Platform, ongoing regional climatic changes seem implied, although the exact nature of such changes cannot be determined from the few available exposure.

Unlike the important environmental changes indicated by the major fills of "Derived Gravels 1A", the paleo-environmental significance of the final rubefaction can be overestimated. The creation of solution cavities in the calcreted surface of Unit 2 argues for reversal in chemical weathering, with solution outpacing calcification, but the difference in climatic parameters may have been small. Despite statements to the contrary by all earlier reporters of the "Red Older Gravels", textural and lithological differences between the rubefied, decalcified gravels ("1A") and adjoining calcreted gravels ("1B") are minimal, and the rubefied gravels are only slightly smaller and better rounded.[4]

Also present on the Holpan Platform, although not present at Holpan I, is a more recent calcrete, postdating rubefaction of "Derived Gravels 1A". This variably textured calcrete is confined to small depressions where fine sands and silty clays have been deposited by the ephemeral drainage. Cementation by Ca and Si can be attributed to evaporation of overland-flow and throughflow waters. This calcrete is probably of upper Pleistocene age.

Unfortunately, dating these events on the Holpan Platform has not yet proved possible (contra Partridge and Brink, 1967). No diagnostic fossils or artifacts have been found, let alone correlated with the deposits at Holpan I. Obviously the "Alluvial Gravels" are very ancient and significantly predate the Younger Gravels. However, the later cut-and-fill events may correlate with some of the younger Older Gravels or even with Younger Gravels events.

Correlating the Holpan I exposure with other deposits on the Holpan Platform is difficult, due to the limited number of deep gravel exposures and destruction of most of the sedimentary record through mining activity during the last 100 years. Another important problem is the limited and infrequent preservation of "Derived Gravels and Sands 2", which results in superposition of "Derived

---

[4]This rebefaction in a lithosol high in ferromagnesian minerals is not to be confused with the intensive, deep-weathering common in perhumid environments and characterized by soils traditionally labeled as laterites and latosols. The deeply-weathered, mid-Pleistocene paleosols along the south coast of Cape Province differ both in kind and degree (see Helgren and Butzer, 1977).

Gravels 1A" and "1B" over the nearly identical "Alluvial Gravels". In addition, "Derived Gravels 1A" and "1B" may also rest directly over the bedrock, so that recognition of these largely colluvial fills is difficult.

The problem of stratigraphic identification becomes evident in evaluating the only, apparently primary, archeologic occurrence discovered on the Holpan Platform. This site, named Suurkree, was excavated by R.J. Mason in 1961 (Mason, 1967). In the upper 60 cm of "Red Older Gravel" as well as in the top of a calcrete layer just beneath it, Mason recovered a small collection of artifacts that he types as Middle and Late Stone Age. Below the calcrete he encountered a massive, calcreted conglomerate. At 75 cm below the conglomerate surface Mason found a single, small, "handaxe-like object". Quite apart from the possibility that the handaxe-like object may be the product of natural fracture[5], it is unclear which deposit at Holpan I correlates with the basal conglomerate at Suurkree. The Suurkree pit has been refilled and could not be re-examined.

Proksch Koppie Gravels

The next lower surface near Windsorton is the Proksch Koppie Platform. This late Paleozoic surface is found between Windsorton and south of Barkly West, and again near Gong-Gong, where it is distinctively developed, at a lower level (see Chap. 3 above). Thin, surficial gravels (also traditionally known as "Red Older

---

[5]Flaked stones similar to the "handaxe-like object" from Suurkree occur rarely in digger's waste heaps on the Holpan and Proksch Koppie Platforms, as well as the Wedburg Terrace (see Fock, 1960). It appears that these flaked stones record a grade of hominid technical ability preceding the Early Stone Age, but presumably postdating primary aggradation of the Wedburg Terrace.

Gravels") subcontinuously litter this platform as well as the gentle slopes below the Holpan Platform. On the surfaces between these gravel deposits are both calcretes and rubefied lithosols. Most of the surficial gravels have been repeatedly redistributed by colluvial processes. Gravel deposits greater than one meter thick are rare, and deposits greater than three meters thick appear to be found in only three areas. The major area of thick gravels is immediately west of the Windsorton townlands, where deep gravels are found on the lip of the Proksch Koppie Platform. These deposits probably demarcate an authentic, but local, late-Cenozoic terrace, coincident with a break in the late-Paleozoic slope. Another thick gravel deposit is found on Harrisdale, 15 km south of Windsorton and also above the right bank of the Vaal. The last area with a thick gravel deposit is "high" Pniel, on the plateau above the left bank of the Vaal, three kilometers south of Barkly West (Fig. 4-3). It is unclear whether these three deep deposits are strictly, temporally comparable, but they do occur at regularly decreasing absolute elevations of about 1137 m, 1127 m, and 1122 m respectively. However, since the modern Vaal gradient is slightly steeper along this now 32 km distance, the gravels at Proksch Koppie are at +30 m above the Vaal, while those at Harrisdale are at +36 m and those at Pniel, +39 m.

The deposits at Harrisdale and Pniel provide little stratigraphic information, but do indicate relict channel positions. Both are typical, granule to coarse-cobble conglomerates, with fine to medium sand matrices. These four-to-six meter thick deposits suggest mid-channel bedloads preserved in structural hollows of the bedrock. Both deposits were post-depositionally calcreted, and later still they were partly decalcified and weakly rubefied. The Pniel deposit, apparently reworked in part, is also locally covered by Hutton Sands.

The gravels at Windsorton show greater variation and record impressive post-depositional alteration. Two thick deposits are exposed at Windsorton. The first along the old Barkly West-Windsorton road, where eight meters of calcreted, typically textured gravels are found resting on Dwyka shale. They are also decalcified at the surface and weakly rubefied. Locally preserved near this pit is a younger, one-meter thick calcrete (no. 190), which stratigraphically overlies the gravels, but is in turn covered by a yellowish red, sandy wash derived from Hutton-type sands. The other, and most important exposure is at Proksch Koppie. This exposure provides a holotype for the geomorphic record documented by the 30 m terrace at Windsorton (Figs. 4-8, 4-9; for gravel morphometry, see Appendix B).

PROKSCH KOPPIE

Unit 1. (top) "Slope Wash 1", 0.3 - 0.4 m, massive, crudely stratified, red (2.5 YR) silty sand.

-- abrupt, wavy, unconformable contact on the pitted and corroded surface of unit 2 --

Unit 2. "Slope Wash 2", 2.0 m, white (10 YR 9/2), moderately sorted, strongly cemented crudely stratified, coarse to medium sands, (no. 096). Also included are rare, thin lenses of subrounded to rounded, fine to coarse pebbles. This unit dips at 2° to 12°, about 45°E of N (toward the Vaal).

-- abrupt, wavy unconformable contact dipping at 4° to 9° northeastward toward the Vaal --

Unit 3. "Gravels and Sands", 4.3 m, complexly interbedded sands and gravels. (1) At top is 0.3 m, strongly cemented, white (5Y), planar-crossbedded, coarse sand, with secondary calcite re-crystallization and many corroded quartz clasts (no. 095). (2) Medially there is 1.9 m, strongly-cemented, granule to boulder conglomerate with a matrix

Fig. 4-8. Proksch Koppie exposure; tunnels at base are in the "Alluvial Gravels"; upper tunnel is in the "Gravels and Sands"; at top is "Slope Wash 2". Staff has 10 cm stripes.

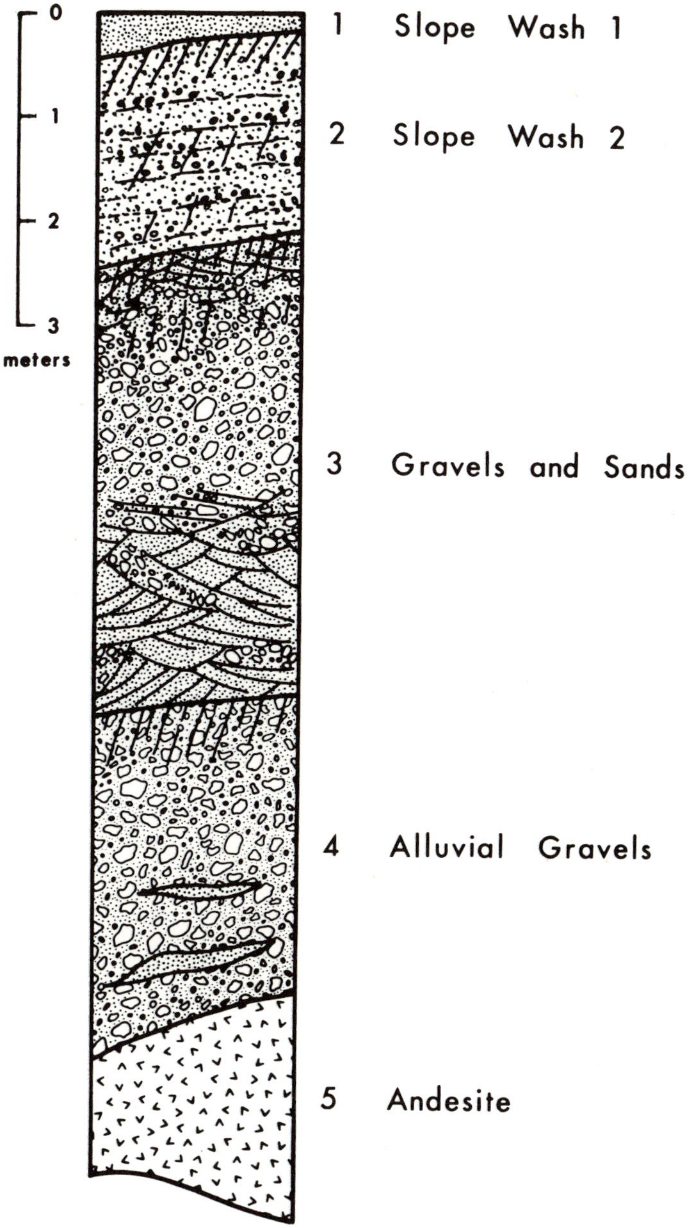

Fig. 4-9. Generalized sedimentary profile at Proksch Koppie.

of medium and fine sand (no. 094, Fig. 4-10). Thin section of the matrix reveals shell fragments and many pseudo-oolite clasts of derived, calcreted, gravels matrix. (3) At the base is a 2.1 m zone where three separate, well-stratified, sand bodies are complexly interbedded with discontinuous, pebble conglomerates. The common, capping sediments are poorly-sorted, light brown (7.5 YR), uncemented, fine and medium sands (no. 093). Predominating below are well-sorted, white (2.5 Y), uncemented medium and coarse sands (no. 091) with 10 cm, internal and capping bands of olive brown (2.5 Y reduced), calcreted, moderately sorted, coarse and medium sands (no. 092).

-- abrupt, wavy, unconformable contact dipping at 2° to 9° northeastward toward the Vaal --

Unit 4. "Alluvial Gravels", >2 m, granule to boulder-grade conglomerate. At the top is a 0.2 m, extremely cemented (with secondary calcite re-crystallization), medium sand (no. 090, Fig. 4-11). The conglomerate below is only weakly-cemented to uncemented, with a matrix of pale yellow (5 Y 8/4, average), very poorly sorted, medium and fine sand (no. 088) and white (5Y), well-sorted, medium sand (no. 089). Included as sand-sized clasts throughout this unit are many corroded quartz grains and calcrete fragments.

-- abrupt, wavy, unconformable contact --

Unit 5. "Bedrock", both dark brown and green Dwyka tillites and Ventersdorp andesite. The Proksch Koppie deposit is preserved to the lee of a glacially streamlined, and striated, whale-back ridge of andesite.

"Slope Wash 1" is a relatively recent deposit, which correlates with a post-Younger Gravels event, probably early Member IV of the Riverton Formation.[6] "Slope Wash 2" and "Gravels and Sands" both

---

[6]This unit passes downslope to the Vaal where it rests on Riverton Member III; see Windsorton I exposure, Chapter 5 below.

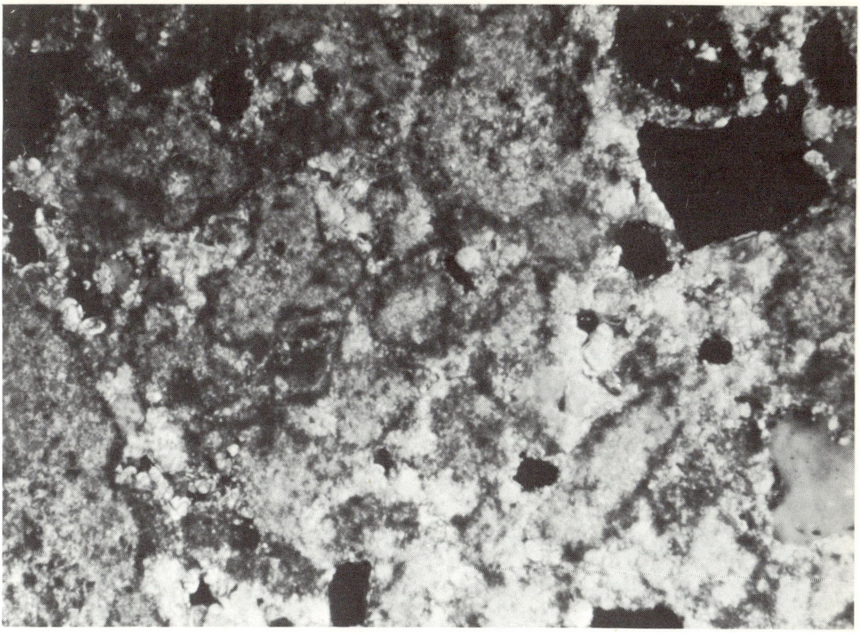

Fig. 4-10. Photomicrograph of sample no. 094, a calcreted gravel matrix in the "Gravels and Sands" at Proksch Koppie. Calcrete clasts and quartz grains are cemented by fine-grained calcite. Frame width is 0.65 mm.

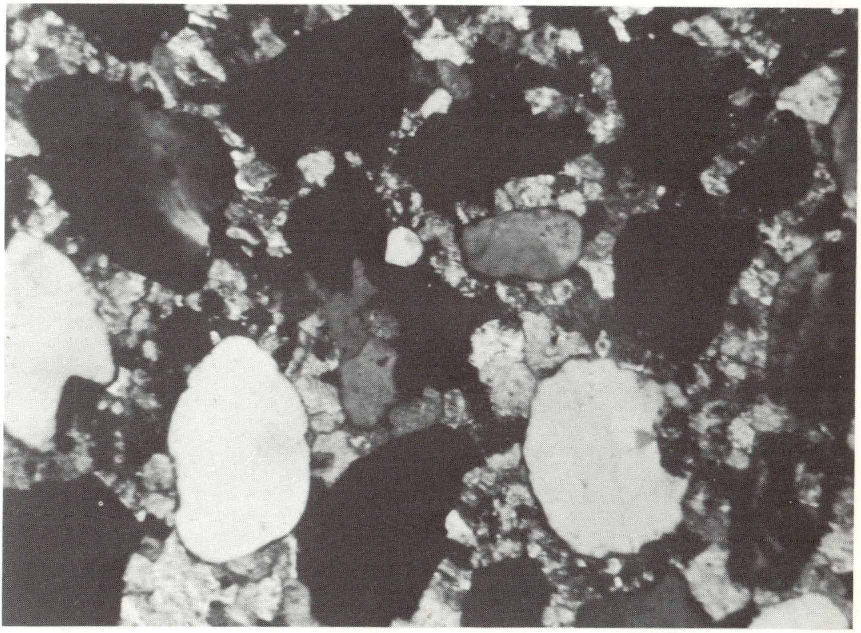

Fig. 4-11. Photomicrograph of sample no. 090, calcreted gravel matrix at the top of the "Alluvial Gravels" at Proksch Koppie. Quartz grains and a K-feldspar cemented by fine-grained calcite. Frame width is 2.0 mm.

are major fills. Thus the Proksch Koppie exposure, like Holpan I, contains two major fill units above authentic basal gravels. Like Holpan I, the Proksch Koppie exposure and its associated deposits reveal an important sequence of fluctuations in the local geomorphic environment. After the major river deposited the basal gravels at Holpan, the stream slid eastward down the gentle pre-Karoo slope. As it cut down through Karoo sediments, irregular exposure of the Holpan and Proksch Koppie Platforms continued. When this down-cutting temporarily stopped, the +30 m terrace, was formed. No primary alluvial deposits apparently remain between the Holpan Platform and the +30 m terrace, and consequently there is no explicit record of river behavior during this interval. The gravels which now litter this area only indicate that the river was at least intermittently capable of mobilizing cobble gravels. However, the total of the erosional processes during this interval were incapable of significantly eroding the pre-Karoo, crystalline bedrock. The "Alluvial Gravels" at the base of the section suggest a typical, mid-channel bedload of a high-competence river. The white sand wedges probably record the primary type of sands carried elsewhere in the channel at this time, while the reddish yellow sands were probably derived from a local bankfall or from a minor tributary. The preservation of calcrete clasts in the gravel matrix indicates that nearby duricrusts of this type were mechanically eroded at this time. A Vaal with high discharges, yet in a semiarid to arid environment is suggested. After the "Alluvial Gravels" were aggraded, they were erosionally truncated (presumably as the channel axis moved away from the site), and the surface calcreted. This calcretion has only a limited downward extent, suggesting this crust to be the result of evaporation of through-flow waters, or ponded flood waters, rather than pedogenesis at this site.[7] The "Alluvial Gravels" reach the surface of the +30 m terrace elsewhere in the Proksch Koppie area.

After the capping crust on the "Alluvial Gravels" developed, the "Gravels and Sands" were aggraded, suggesting a channel-margin flood deposit as the Vaal channel axis briefly returned to the site. At the base of the unit are complex, sandy interbeds, perhaps produced by mixing of mid-channel sands (white?), flood-basin sands (olive brown?) and bankfall or lateral-creek sands (light brown?). Above the sandy interbeds are gravels of a mid-channel traction-load along the channel axis. The preservation of fragile shell fragments in the gravel matrix collaborates the interpretation that the aggradation of this unit occurred during a brief, high-magnitude discharge, while incorporation of clasts of previously calcreted gravel matrix provides additional evidence of a major unconformity between this unit and the basal gravels. At the top of Unit 3 are cross-bedded sands formed by local dune movement on the channel bed. These sands indicate the channel axis again migrated away from the site, perhaps as the flood surge passed. The "Gravels and Sands" were then eroded and probably calcreted, (perhaps as the Vaal incised down to the Wedburg Terrace level), creating another, major erosional unconformity. "Slope Wash 2", composed of crudely-bedded sands and rare gravels, was deposited at a later date perhaps correlating to a Younger Gravels phase. This unit indicates a period of general mobilization of surficial sediments from upslope, implying a discontinuous ground cover. Later, this unit and the upper parts of the "Gravels and Sands" beneath it were "calcreted". Much later, following weathering of the surface

---

[7] In a surficial sediment suite such as at Proksch Koppie, the high proportion of weathered quartz grains probably is a progressive post-depositional alteration, and only provides a general index of antiquity.

of Unit 2, "Slope Wash 1" was deposited to cap the section. This unit also indicates sediment mobilization upslope.

Correlating the Proksch Koppie exposure with the two-kilometer distant exposure at the Barkly West road is difficult. The gravels at the latter site may coincide with the "Alluvial Gravels" or the "Gravels and Sands". Or the gravels at the Barkly West road may record an alluvial aggradation on the +30 m terrace not preserved at Proksch Koppie. The rubefaction of the gravels along the Barkly West road probably provides little additional evidence, but it is compatible with the corrosion of the surface of Unit 2 at Proksch Koppie. Since the calcreted horizons at Proksch Koppie are largely the product of precipitation from throughflow waters, cementation at a comparable, but better-drained site can be expected to be more variable.

The environmental fluctuations suggested by the Proksch Koppie exposure are in the magnitude of those identified for the late Pleistocene record of the region (Butzer, Helgren, _et al_., 1973; Butzer, Fock, _et al_., 1973). The most profound fluctuations are those between calcrete genesis and calcrete corrosion, and between stability and mobility of the surficial sediments. The latter is analogous to the erosion caused by modern overgrazing but, of course, a fluctuation in vegetation patterns related to climatic change must be postulated for the Proksch Koppie stage. While the modern Vaal carries very few coarse cobbles in its load,[8] the coarse gravels at Proksch Koppie (as at Holpan) may not be environmentally important. Flood velocities of the modern Vaal are sufficient to mobilize comparable gravels, and the Vaal during the

---

[8]The general absence of coarse clasts in the modern load of the Vaal is directly related to the "cushion" of fine-grained, Riverton sediments which provides few gravels when modern bank erosion occurs.

Proksch Koppie stage may have had a larger basin upstream, or simply different channel morphologies in the Windsorton area.

There have been no systematic archeologic investigations of the gravels remaining on the Proksch Koppie Platform. No fossils have been recovered, but van Riet Lowe (1952: 8) reported that "rolled" "Kafuan and Olduwan" tools occur in diggers spoil heaps on the surface. Whether the manufacture of these tools is coincident with the primary gravel aggradation on this surface is unknown, but doubtful. The date of the initial aggradation of the Proksch Koppie gravels is also unknown, and a "late Tertiary" appellation may be tentatively applied.

Droogeveld Gravels

The next lower planation feature near Windsorton is the Wedburg Terrace, which provides a stratigraphic benchmark throughout the Lower Vaal Basin. Elsewhere along the Vaal there are a variety of other Older Gravel deposits which, like those on the Vaal-Harts interfluve, also occur above the Wedburg Terrace. Foremost among these are the deposits at Droogeveld.

Located 14 km south of the Vaal-Harts confluence and at a maximum elevation of 1135 m, the Droogeveld gravels are irregularly spread across the farms Mozib, Than, and Droogeveld (all now part of Sydney) and adjoining parts of Pniel and Klipfontein (now the northeast part of Rooipoort) (see Fig. 4-12). The gravels litter slopes of the Pre-Karoo Surface and are not associated with planation features. Diamond digging near the turn of the century removed most of the primary deposits and all of the old pits have been at least partly refilled.[9] Apparently most of the deposits consisted

---

[9]DeBeers Consolidated Mines re-examined the Droogeveld deposits in 1968. Their pits have been largely refilled.

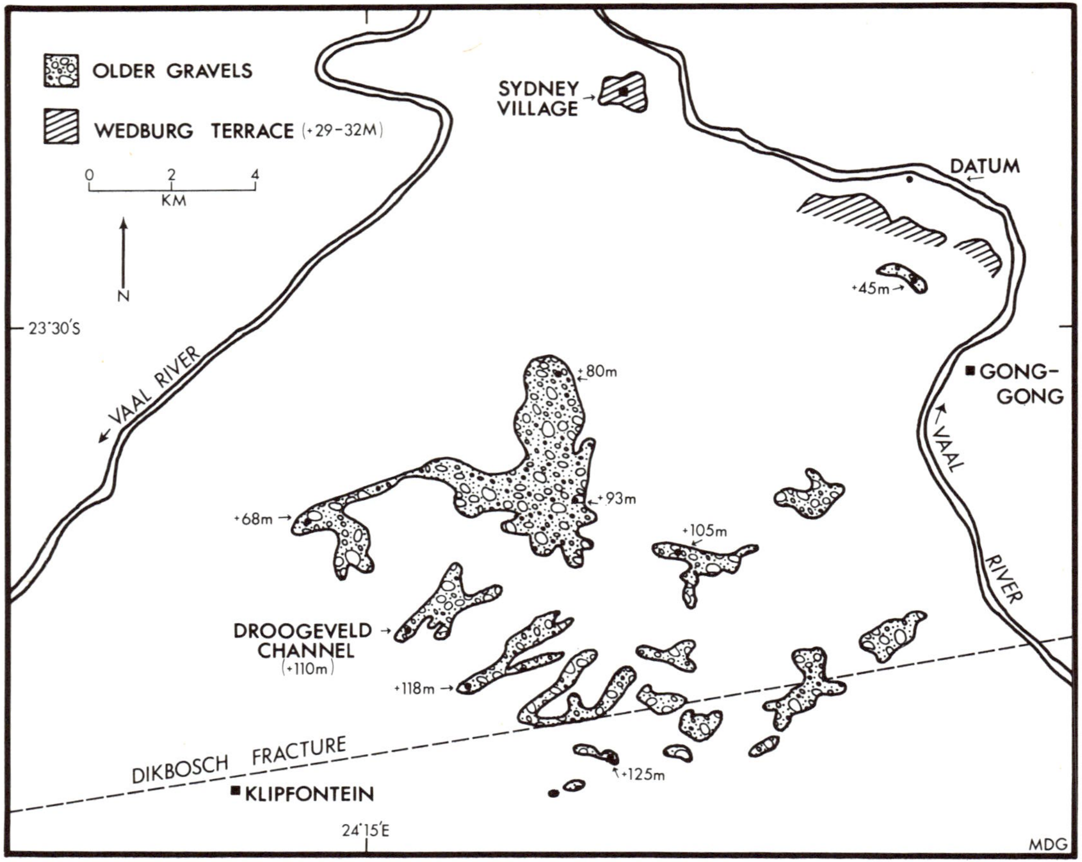

Fig. 4-12. Older Gravels deposits and the Wedburg Terrace near Droogeveld.

of a layer of gravels about 30 cm thick beneath a meter or more of red Hutton Sands (see Wagner, 1914: 267-270).

When the total spatial array of the Droogeveld deposits is considered, no systematic pattern is apparent. However, deposits deeper than two meters are confined to long, linear depressions seldom more than 10 m wide and 5 m deep. These linear depressions coincide with the trace of the Dikbosch Fracture and the lineaments (master joints?) which radiate outward from it. This patterning indicates these late-Cenozoic gravels are colluvial fills in structural depressions in the Pre-Karoo Surface.

At Droogeveld the strike of the Dikbosch Fracture is marked by a shallow valley, 0.5 to 1.0 km wide. This valley has been masked with one to five meters of red, Hutton Sands, and the bedrock profile in the valley is obscured. The lineaments radiating from the fracture are indicated by even less impressive depressions and may be entirely filled with red sands. At Droogeveld the Dikbosch Fracture strikes at 82° east of north, and is locally known as "Coqui's Gulley". The most important diamond-bearing deposit was in the "Droogeveld Channel", which strikes at 48° east of north and intersects the fracture at an angle of 34°. Other lineaments tend to run subparallel to these principal axes.

Fortunately, one deep exposure of the gravels in the "Droogeveld Channel" still exists (Figs. 4-13, 4-14; gravel morphometry data appear in Appendix B).

DROOGEVELD CHANNEL

Unit 1. "Hutton Sands", 2.0 to 2.5 m, red (2.5 YR 5/8-10), well-sorted, vertically structured, fine sand (no. 139). Like all red sands in the region, these are composed of quartz, with surficial, iron-oxide stains.

-- abrupt, wavy, unconformable contact --

Fig. 4-13. Exposure of the Droogeveld Gravels; vertically structured Hutton Sands veneer the thin "Fine Gravel" horizon and the "Coarse Gravels" near the hammer.

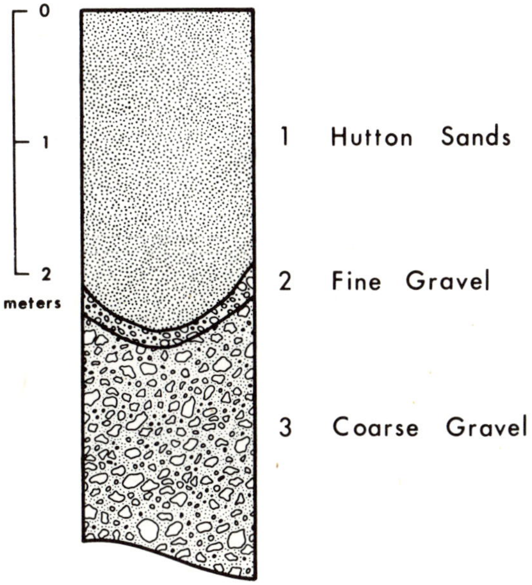

Fig. 4-14. Generalized sedimentary profile at Droogeveld.

Unit 2. "Fine Gravel", 0.1 m, granule to medium-pebble gravel with a matrix of red (2.5 YR 4-5/6-8), very poorly sorted, fine sand (no. 140). The fine gravels are angular to rounded, strongly-weathered, chert, andesite, quartz, and quartzite.

-- abrupt, wavy, unconformable contact --

Unit 3. "Coarse Gravels", more than 2 m (base unseen), strongly-weathered, uncemented, granule to coarse-cobble conglomerate of angular to rounded, silica-rich rocks (quartz, quartzite, chert, agates, etc.) mixed with angular and sub-angular, fractured and rotting andesite. The gravel matrix is red (2.5 YR 4/8), very poorly sorted, fine sand (no. 141).

This sequence of a basal gravel succeeded upward by a thin band of fine gravel and a massive sand unit closely parrallels the sections illustrated by Wagner (1914: 267-270), who saw exposures while primary digging was underway. The dominant bedrocks of the Droogeveld area are Pniel Series volcanics, but thin veneers of Dwyka sediments are still preserved. The bedrock profiles in the Droogeveld Channel described by Wagner show them to be crudely rectangular, and to have greater depth than width. Wagner also noted that the andesite at the channel margins was highly fractured, resulting in many boulder-sized fragments. Wagner's sections include two layers which do not occur in the section above. One layer described by Wagner was "Coarse detritus with much angular matter", irregularly found above (and below?) the "Fine Gravel", which Wagner labeled "Kaffir Corn Layer". Wagner identified his "Coarse detritus" as a "rain wash" deposit, partly mixed with the upper "eolian" sands. Wagner's other layer, at the base of the section was a thin horizon of "bantams" -- the heavy, well-rounded, schist and gneiss pebbles associated with rich, diamond deposits.

While a coarse detritus layer at the base of the "Red Sand" is readily believable, Wagner's variations below the "Fine Gravel" probably represent several facies of the "Coarse Gravels".[10]

The "Coarse Gravels" are not a primary alluvial deposit, but colluvial gravels derived from an alluvial deposit no longer preserved (as suggested by Cooke, 1947, contra Wagner, 1914). Consequently the depositional record of the Droogeveld Gravels is necessarily incomplete. The sequence begins with erosion of the cover of Karoo sediments. At this time the fracture zones along the lineaments of pre-Karoo bedrock were also eroded and widened. The "Coarse Gravels" were then colluvially dumped into linear depressions as part of a general mobilization of surficial sediments down the pre-Karoo slopes, long after the river had abandoned the Droogeveld area. At a later date the "Fine Gravels", best viewed as a weathered and abraded residual of the "Coarse Gravels", were deposited. And at a still later date the mainly eolian "Hutton Sands" was deposited and probably accompanied, at least initially, by another interval of increased colluvial activity.

There is then no direct record of river behavior in the Droogeveld area. As noted above (Chap. 3), Cooke (1947) speculated that the Droogeveld course of the Vaal was abandoned after a capture by a tributary of the Harts. However, this theory is not substantiated by the Droogeveld deposits, which are spread at progressively lower elevations. These deposits suggest that the Vaal gradually

---

[10]The bantam layers (placers) identified by modern diamond diggers along the lower Vaal always have proved to be of a lower stratigraphic category than the fill units identified here. Placer horizons typically are discontinuous in exposure, and are located most commonly at the base of major fill units.

slid northward down the adjacent Karoo slopes to re-establish the Paleozoic drainage line marked by the modern bend at the Harts River confluence.[11]

No artifacts or fossils have been reported from the Droogeveld Gravels. There is no way to explicitly correlate the Droogeveld Gravels with other deposits in the Lower Vaal Basin. On altimetric grounds they could correlate with any or all of the Nooitgedacht, Holpan, or Proksch Koppie gravels. However, the lithology and weathering characteristics are most similar to the Nooitgedacht gravels.

The Geduld Bench

Another high-level surface with associated gravel deposits is the Geduld Bench, found at two localities, both west of the Vaal-Orange in front of the Gaap Escarpment. The largest occurrence is southwest of Schmidtsdrif (at 28° 50' S., 23° 50' E.) at an elevation of 1100 m, about 110 m above the Vaal. The other locality is below the Orange-Vaal confluence (at 29° 08' S., 23° 30' E.), at an elevation of 1090 m, locally 120 m above the Orange (Fig. 4-19).

The genesis of the Geduld Bench, as well as its significance for the regional denudation record is problematic. At both localities Transvaal System sedimentaries with a thin veneer of Dwyka

---

[11]High level gravels also occur on Klipfontein and Volstruispan (both now part of the farm Rooipoort). None of these deposits have ever been mined. However, these deposits do not trace a past Vaal channelway (contra Wagner, 1914). The Older Gravels deposits along the Bushmanskloof Gorge course of the Vaal indicate great longevity for this channelway. The Klipfontein gravels probably have a history like those at Droogeveld, except that they migrated down a south-facing, pre-Karoo slope. The history of the gravels on Volstruispan is less clear. But they are located below hills in the Pre-Karoo Surface, still mantled with Dwyka tillite. Pending excavation, it must be assumed that these gravels are flootslope deposits derived from tillite.

tillite are irregularly exposed at the surface. Small gravel bodies are located along the eastern, "outer" edge of the bench, adjacent to tillite horizons. The 1 to 2 m deep gravels are strongly cemented (no. 212) and composed of the typical, inhomogeneous, granule-to-cobble grade clasts. In the available exposures the gravels contain no interbeds and do not exhibit depositional structures diagnostic of aggradation by a major river. The location of the gravel deposits on or near tillite horizons suggests that the gravels are derived directly from local weathering and erosion of tillites. Since the Geduld Bench does not bear diagnostic alluvial deposits, it cannot be definitely identified as a late-Cenozoic alluvial terrace or platform. In addition to the alluvial terrace explanation, other possible origins of the Geduld Bench must be considered. This feature may be (a) an exhumed, structural bench in the pre-Karoo bedrock, with a thin tillite coating, (b) an exhumed planation feature of the Pre-Karoo Surface, or (c) a low-angle, pediment-form plain beginning behind a resistant bedrock threshold of the Pre-Karoo Surface. Unfortunately there are only a limited number of brdrock exposures and none of these possibilities can be eliminated. The Geduld Bench will remain an enigmatic landform until more diagnostic exposures are either discovered or created.

No artifacts or fossils occur in the Geduld gravels. Any correlation of these deposits with others in the lower Vaal River Basin is tenuous.

Other Older Gravels

There are several additional Older Gravels deposits in the Lower Vaal Basin, which must also be reported. These occur in two general areas, one upstream near Warrenton, and the other midstream near Schmidtsdrif.

At Warrenton a structural bench of Ventersdorp volcanics is found +21 to +23 m above both banks of the Vaal (1190 m absolute elevation) (Fig. 4-15). On the south (left) bank, this bench is coated with typical "Red Older Gravels". While no primary deposits remain, it is apparent from the digger's pits that the gravels were seldom more than a meter thick and probably did not exhibit primary depositional structures. The Warrenton gravels are paired with a deposit located west of Warrenton, at Sydney's Hope (elevation 1185 m). Here, situated +23 m above the Vaal is a thick, massive calcrete, presumably recording an ancient, fine-grained, alluvial deposit. This strongly-cemented calcrete is 3.5 m thick and includes Dwyka shale in its base. The non-HCl soluble components are silty clay sands and clayey silt sands (nos. 192, 193). This deposit exhibits extensive, secondary and tertiary calcite recrystallization, which has destroyed any primary bedding structures that may have been present.

Small patches of high-level gravels occur sporadically between Warrenton and Windsorton. These are the typical, thin, rubefied and weathered gravels, sometimes calcreted of depth and which have been excavated and resorted by diamond diggers. Across from Witrand such gravels occur at +24 to 26 m above the Vaal (1170 m), where they litter the upper valley slopes. Further downstream gravels occur near Spence's Kop and Waterval at about 1160 m, +29 m and +31 m above the Vaal respectively. Less ancient gravels locally fill tributary gullies. These other high-level gravels all exhibit extensive colluvial redistribution. Whether these Older Gravels deposits below Warrenton correlate with each other is unclear, and suggesting correlations with more distant deposits is even more hazardous. On altimetric grounds the gravels on the Warrenton townlands could correlate with any or all of the Older Gravels deposits near Windsorton.

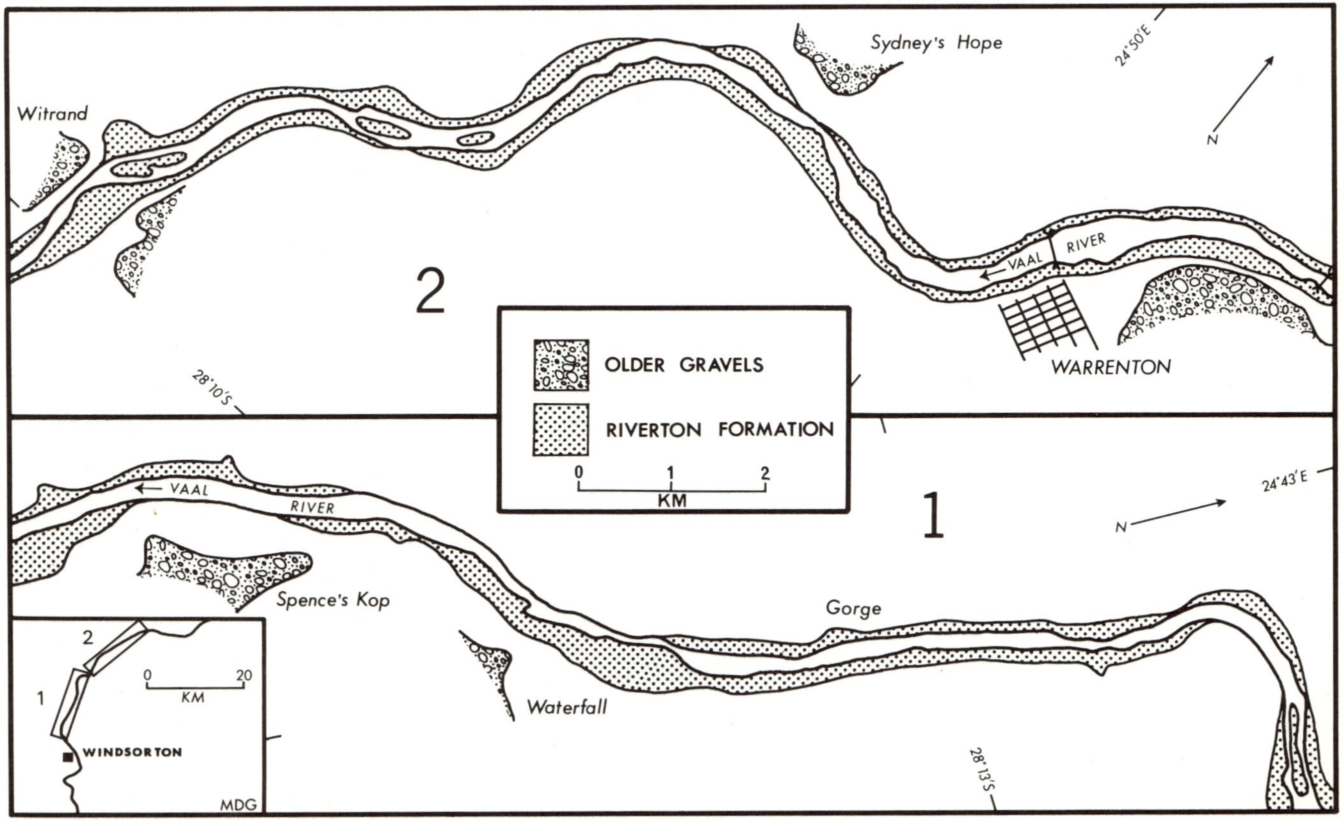

Fig. 4-15. Older Gravels features and the Riverton Formation between Warrenton and Windsorton.

The Older Gravels north of Schmidtsdrif are also spread across the high flanks of the Vaal valley. They first appear near Muirton and extend southward to Nuwe Hoop (Fig. 5-31). In all cases the deposits are typical, shallow, colluvially reworked veneers of rubefied gravels, which in many areas occupy solution pits in calcreted gravels. No primary alluvial structures are preserved. Many deposits are located near tillite outcrops. The general elevation is 1075 m, about 75 m above the Vaal. This is about the same level as that of the mesa crest opposite Schmidtsdrif and the unexposed gravels on Volstruispan. A late-Cenozoic terrace level may be demarcated and since these gravels occur above the Wedburg Terrace, it is tempting to correlate them with the +30 m terrace (Proksch Koppie) at Windsorton. However, such a correlation cannot be substantiated by other evidence.

The origin and significance of the Older Gravels deposits near Warrenton and Schmidtsdrif is even more obscure than the previously described examples. It can not be determined from the available exposures whether they record significant ancient floodplain elevations, or whether they are little more than colluvia conviently deposited on flat terrains of the Pre-Karoo Surface.

The Wedburg Terrace

The Wedburg Terrace forms a morpho-stratigraphic marker throughout the Lower Vaal Basin. It is well-developed along the Vaal between Windsorton and Barkly West and again between Gong-Gong and Douglas. In addition, parallel terraces occur on the Riet, Harts, and Orange Rivers.

At Windsorton the Wedburg Terrace occurs on both sides of the river (Fig. 4-16). On the west bank it is developed in Dwyka shale, both north and south of the Windsorton townlands. On the east bank it is developed in Dwyka shale at Wedburg, and it is

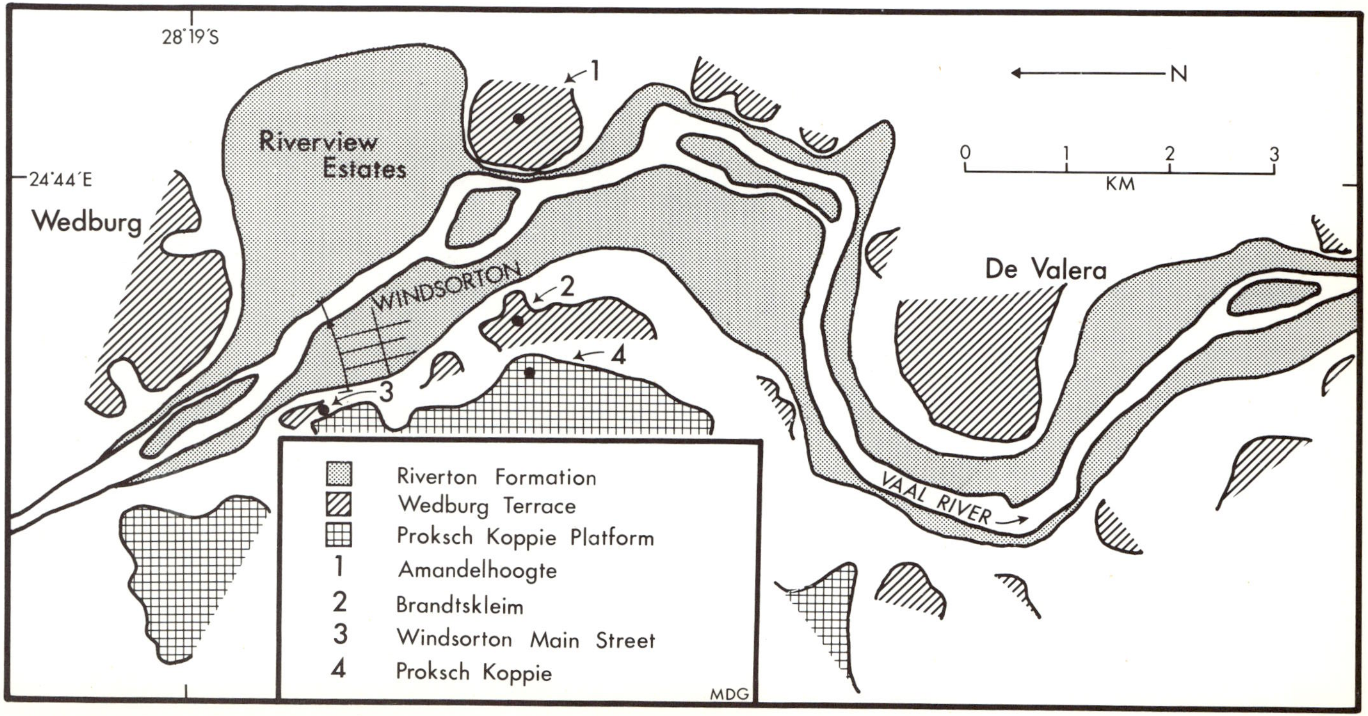

Fig. 4-16. Wedburg Terrace and Proksch Koppie surface near Windsorton.

coincident with a shale and tillite-coated bench of andesite at Amandelhoogte.[12] These features are at an absolute elevation of about 1130 m, i.e. +22 to +24 m above the Vaal.

A deep gravel exposure was described at Wedburg by Wagner in 1914 (pp. 264-265), so that this toponymn can be formally applied to this morpho-stratigraphic unit (Fig. 4-17). Modern exposures at Wedburg are no longer as deep or comprehensive as witnessed by Wagner. Entirely missing in the modern exposures, but described by Wagner, are the capping 30 cm of red (Hutton-type?) sands and the upper 80 cm of diamond-rich, weathered and strongly rubefied gravel. Below these horizons Wagner demarcated two more, less-weathered gravel layers totaling 4 m in thickness. This zone is present in the modern deposit. He identified these layers as separate gravel aggradations. The modern exposures at Wedburg, while only three meters deep, do not support Wagner's interpretation of separate, gravel deposits. They do show one gravel aggradation progressively more weathered from base to top. Matrix colors change from very pale brown (10 YR) at the base to yellow (2.5 Y) in the middle, and red (2.5 YR) at the top (nos. 075, 074, 076). The subangular to well-rounded gravels are composed of granule to boulder-grade andesite and silica-rich rocks (i.e. quartz, quartzite, chert, agates, etc.), with only a minor proportion of locally derived shale. Throughout the deposit, many of the andesite clasts are thoroughly rotted, and readily crumble when handled. At the base of the deposit the uncemented matrices are medium to coarse sands. However, as the intensity of post-depositional weathering increases upwards, the matrix becomes finer, achieving a sandy clay silt texture.

---

[12]Younger Gravels occur on the farms Wedburg and Amandelhoogte. The presence of both Older and Younger Gravels on these farms was not made clear by Söhnge and Visser (1937) and has been subsequently confused by Partridge and Brink (1967).

Fig. 4-17. Vaal River at Windsorton; Wedburg is the vegetated surface in the upper left center. Abandoned diggers' pits in Riverton Formation deposits are in foreground.

Other vestiges of the Wedburg Terrace can be traced almost continuously around Windsorton, e.g. Windsorton's "main street" is at this level, as is an impressive break in slope east of Riverview Estates, as well as a bench below Proksch Koppie (see Fig. 4-16). The only other thick gravel deposit at this level is found at this last locality, here named Brandtskleim.[13] The Brandtskleim exposure may be described as follows (Fig. 4-18).

---

[13]The Brandtskleim area was called "Little Proksch Koppie" by Partridge and Brink (1967).

Fig. 4-18. Exposure at Brandtskleim; hammer is at surface of the "Basal Gravels".

BRANDTSKLEIM

Unit 1. "Slope Wash", up to 0.4 m, crudely stratified, red to yellowish red (2.5 to 5 YR), apedal silty sand.

-- abrupt, irregular, unconformable contact with local, conical solution-cavities --

Unit 2. "Alluvial Sands", 1.0 to 2.0 m in total. The upper 10 to 20 cm is a white (10 YR) to pink (5 YR), indurated calcrete, which passes downward into 1.0 to 1.5 m of pink (5 YR 7/3), variably cemented, well sorted, silty sands (nos. 305, 306), with low angle, planar and trough cross-bedding. At the base is 30 to 80 cm, weakly cemented, moderately sorted, subrounded to rounded, fine-cobble conglomerate (dominately andesite) with lenses of medium and coarse, well-sorted sands with planar and trough cross-bedding. The gravel matrices are very pale brown (10 YR 8/2) to yellowish red (5 YR 5/6), moderately to well sorted coarse sands (nos. 303) and silty sands (no. 303), dominated by derived calcrete and quartz.

-- clear to abrupt, undulating, unconformable contact --

Unit 3. "Basal Gravels", 0.3 - 1.2 m, crudely stratified, subrounded to rounded, pebble conglomerate with dispersed cobbles, and local sand lenses. The matrix, which is calcreted only at the surface, is very pale brown (10 YR 8/3), well sorted, calcrete and quartz sands (no. 302).[14]

-- abrupt, undulating unconformable contact --

Unit 4. "Bedrocks", 0.2 - 0.8 m of green, Dwyka tillite over 0.2 to 0.3 m of dark-brown, Dwyka tillite, veneering a whaleback ridge of Ventersdorp lava.

---

[14]Nearby Brandtskleim Unit 3 intersects the terrace surface and assumes more typical, coarse gravel textures. The matrix is often strongly calcreted.

While the capping colluvium (Unit 1) is reworked Hutton Sands, and compares to Units 1 at Proksch Koppie (above) and Windsorton I (Chapter 5), the Brandtskleim exposure significantly records two thin but separate alluvial fills. Each of these fills was post-depositionally calcreted, and each includes an inhomogeneous array of sorted and bedded sediments, derived from erosion of tillites and mechanical abrasion of nearby calcretes. The textures and bedding are compatible with a high energy Vaal, but preservation of derived calcrete clasts suggests a relatively arid, local environment. The separate calcretion of Units 2 and 3 probably indicates continuing, low-amplitude environmental changes as well as intervals when the Vaal channel axis migrated away from this channel-margin site. Brandtskleim is the only exposure where two intervals of alluviation are recorded for Wedburg Terrace-times. No doubt only fragments of the original Wedburg deposits survived the strong erosional interval at the beginning of Younger Gravels time.[15]

The calcretion of Brandtskleim again is the product of precipitation from throughflow waters or ponded floodwaters. These calcic sediments are notably different from the rubefied and weathered sediments across the river at Wedburg. A first explanation would be that deposition and weathering at Wedburg preceded or followed all the events at Brandtskleim. However, these differences are probably partly the product of different local settings. The gravels at Wedburg are on a freely drained site, not influenced by throughflow waters from upslope. Consequently the formation of solution cavities on the surface of Unit 2 at Brandtskleim may fully parallel the rubefaction and weathering at Wedburg.

---

[15]The Wedburg Terrace residuals are so impressive around Windsorton, that it is not unreasonable to speculate that a sediment pool as large and complex as that of the combined Younger Gravels and Riverton Formation may once have been present.

Between Windsorton and Barkly West, the Wedburg Terrace is irregularly preserved on both sides of the river (Figs. 4-16, 5-16). South of Amandelhoogte it appears on another andesite spur at Devalera, and still further south on the east bank it is masked by Hutton Sands. However, the most impressive development of the Wedburg Terrace in this reach is along the west bank at Slangheuvel (also at Harrisdale and Willowbank; see Fig. 2-18). Here the margin of the Wedburg Terrace is marked by a long scarp of green and blue-green Dwyka shales, with a calcreted, flat summit at +28 to 29 m above the Vaal. Irregularly cemented into the one to two meter-thick calcrete are colluvially reworked gravels. Diggers waste heaps suggest that some of these deposits may have been one meter deep, but no primary deposits remain. Typical, Older Gravel solution pits occur on the surface.

Along the steep reach of the Vaal below Barkly West reworked, colluvial washes of Older Gravels litter the high surfaces and slopes leading into the river valley. The Wedburg Terrace reappears above the Vaal's left bank between Gong-Gong and Longlands, where it stands at +26 to 29 m (Fig. 4-12). The deposits consist of calcreted gravels up to 1 m thick, buried beneath Hutton Sands. Where exposed, the calcreted surface is marked by typical, conical solution cavities. No primary alluvial deposits appear to be preserved.[16]

Below Delportshoop, at Muirton, the Vaal enters another, incised pre-Karoo channelway -- the Bushmanskloof Gorge. Dwyka sediments along the Vaal channel are 140 m and more below ridge-crests of Ventersdorp lavas. As the Bushmanskloof Gorge opens at Nuwe Hoop, the Wedburg Terrace reappears as a calcreted gravel surface cut in a Dwyka dropstone bed at +33 to 34 m (absolute elevation about 1030 m, Fig. 5-31). This gravel is up to six

---

[16]The bedrock in this area is Ventersdorp andesite and this platform may be related to pre-Karoo erosion.

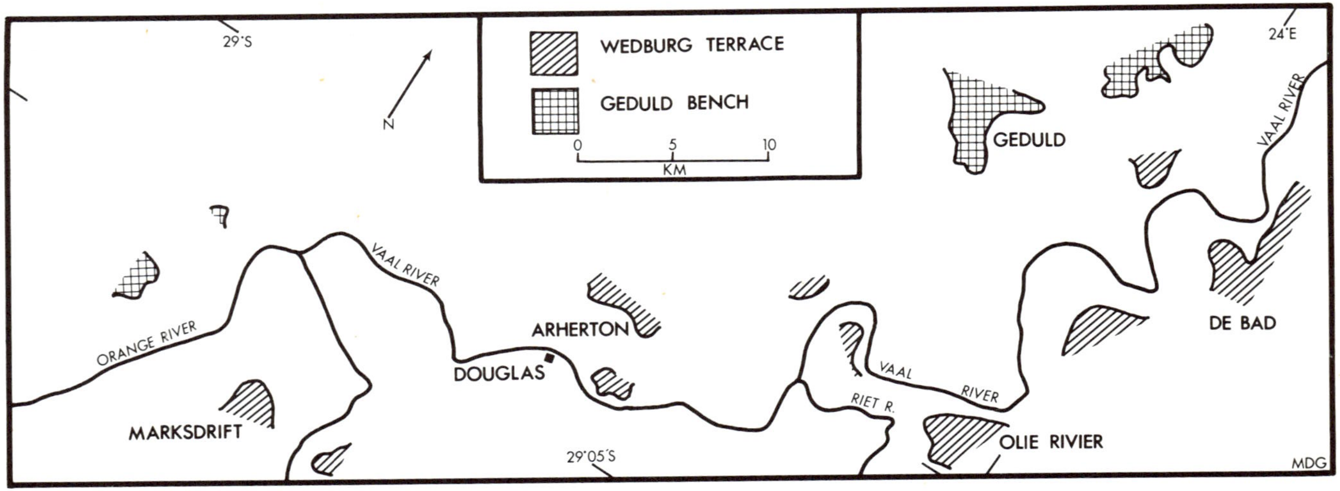

Fig. 4-19. Wedburg Terrace and Geduld Bench -- DeBad to Read's Drift.

meters thick and composed of the typical, subangular to well-rounded, granule-to-boulder grade clasts of andesite, silica-rich rocks, and locally derived shale (gravel analysis in Appendix B). The calcreted matrices are light gray (5 Y), clayey silt sands (no. 138).

Between Schmidtsdrif and Douglas, a distance of over 50 km, planar surfaces or hillslope inflections representing the Wedburg Terrace are nearly continuous along both the east and west sides of the river (Fig. 4-19). West of Schmidtsdrif a broad bench cut into Transvaal System sedimentaries and with 1 or 2 m veneers of calcreted gravels occurs at +32 to 34 m. Further south, on the west bank, calcreted gravels occur on Sivonel and Platdrif at +32 to 34 m. However, the most impressive occurrences of the Wedburg Terrace are along the east bank at Middelplaats North, De Bad, and Olie River. Here, long (2 to 5 km), low angle (less than 1°), pediment-form footslopes are truncated by a steep scarp that rises +31 to 37 m above the Vaal. This bench is developed in Dwyka shale, and is typically capped by one or two meters of calcrete, with gravel inclusions, beneath subdued dunes of vegetated Hutton Sands. At Middelplaats North some alluvium of the Wedburg Terrace is still preserved (at +31 m), and can be described as follows (see Fig. 5-31).

<u>MIDDELPLAATS NORTH</u>

Unit 1. "Slope Wash", up to 20 cm, strong brown (7.5 YR 5/5), humic silty sand (no. 281) with mechanically abraded (derived), subrounded to rounded, fine pebbles of calcrete.

-- abrupt, smooth disconformity --

Unit 2. "Calcrete", 1.0 to 1.2 m indurated, silty sands with colors in the pinkish gray range (7.5 YR 7/2-3, nos. 282, 283) which have been weathered at the surface into angular, cobble-grade (rhomboidal) blocks. The 1 to 5 cm wide, interblock fissures near the surface are filled with silty sands from Unit 1. At depth fine, horizontal stratification is well developed, on which weak, vertical structuring is imprinted. The calcrete preserves secondary and tertiary calcite crystallization. The unit is a thick, calcreted, channel-margin vertisol.

-- clear, smooth, unconformable contact --

Unit 3. "Alluvium", the upper 1.3 to 1.8 m is reddish yellow (7.5 YR 7/5), horizontally stratified, silty sand cemented (indurated) by amorphous silica and calcite (no. 284). The lower part is greater than 1 m, reddish yellow (7.5 YR 7/6), horizontally stratified, silty sand, which is indurated by calcite (no. 285). Kaolinitic clays occur throughout.

Preserved at Middelplaats North is a thick sequence of arenaceous, alluvial Wedburg Terrace sediments, probably subjected to multiple episodes of post-depositional cementation and weathering. They are capped by humified, colluvially reworked Hutton Sands. Bedrock is not exposed at Middelplaats North, but Black Reef sedimentaries, with veneers of Dwyka shale are present nearby. The sand fractions of Units 2 and 3 are dominated by quartz and fragments derived from Black Reef sedimentaries. About 200 m north of Middelplaats North exposure, calcreted gravel veneers mark the Wedburg Terrace level. Together these deposits suggest a wide, sandy Vaal channel with local, coarse-gravel splays.

Above Douglas the Wedburg Terrace is well-developed along the west side of the Vaal on Arherton. Adjacent to the Vaal a glacially shaped knob of Ventersdorp lavas bears a thin coat of calcreted gravels at +38 to 39 m, and 4 km north a retreating scarp of calcreted Dwyka shale stands at +40 to 41 m.[17]

---

[17]A comparable slope inflection may be buried beneath Hutton Sands south of Douglas.

Despite the ubiquity of the Wedburg Terrace, informative coeval deposits are restricted to those at Brandtskleim and Middelplaats North, while veneers of calcrete (largely pedogenetic) and sporadic, often reworked gravels are more typical. Furthermore, there is no guarantee that the deposits reported here are strictly, temporally comparable. The Wedburg Terrace deposits clearly indicate that the Vaal was at least locally (and probably repeatedly) capable of carrying boulder-grade gravels, but data on channel morphologies is absent and the accompanying fine sediments suggest a river not unlike the present. However, the Wedburg Terrace does demarcate a major stage in the Cenozoic denudation of the region. After abandoning the +30 m terrace at Windsorton, the Vaal cut down through Dwyka sediments and stopped long enough at this level for the extended shale plains at Slangheuvel and De Bad to be eroded. Subsequently the Vaal incised these surfaces throughout the region to create the Younger Gravels channelway.

No artifacts have been explicitly reported from the deposits on the Wedburg Terrace. However, Van Riet Lowe (1952) implies that rolled, pre-Acheulian artifacts were recovered from diamond diggers wasteheaps <u>and</u> the terrace surface. No Acheulian artifacts, either <u>in situ</u> or in waste heaps, were found by this investigator in the Wedburg Terrace deposits. Consequently, the Wedburg Terrace is assumed to be pre-Acheulian in age, although potentially coincident with a pre-Acheulian technical grade.

Sources of the Vaal River Gravels

Sources of the large clasts in the surficial conglomerates of the Lower Vaal Basin are an important index of the nature of regional, Cenozoic denudation. Clasts in sedimentary deposits are produced by subaerial weathering and erosion of available bedrocks, while mixing, sorting, and abrasion are effected by the subsequent transport media. Coarse clasts can be produced by a variety of settings. There appears to have been three primary sources for the Vaal River gravels.

Dwyka tillites and dropstone beds were no doubt the most important, primary source, providing "ready made", gravel-sized particles, as explicitly recognized by du Toit (1907) and Wagner (1914). The clasts from the alluvial and glacial deposits are highly similar, having comparable sizes and shapes. Table 4-1 compares the morphometry of clasts from three tillite exposures and two alluvial gravels (see also Appendix B). In addition, gravel clast lithologies compare closely with those of nearby tillites.[18] The only notable differences are better rounding and slight size reductions, as can be expected, in the alluvial deposits. The few, Karoo-age clasts (shale, dolerite, fossil wood) in the gravels are of course missing from the tillites. All the andesite clasts could have been introduced into the gravel deposits via tillites and no cutting or plucking of pre-Karoo bedrock was necessary to provide these clasts.

Another primary, but less significant source for coarse clasts are the fine gravels dispersed in the bed-loads of nearly all rivers on the high plateaus of southern Africa. These pebble-grade particles of quartz, agates, chert, heavy mineral schists and gneisses and quartzite, produced by erosion of various Precambrian rocks in the Vaal basin, are virtually indestructible, and probably have migrated intermittently across the subcontinent since the early Mesozoic.

The last primary source for coarse-grade clasts are Karoo rocks. Shale clasts and those of similar tillite matrices occur in the gravel deposits only where thick shale strata are found. Karoo dolerite and indurated shale clasts, while frequently reported by

---

[18]See also Appendix B.

TABLE 4-1

COMPARISON OF CLASTS SHAPES AND SIZES IN THE TILLITES AND THE LATE-CENOZOIC GRAVELS[a]

	L	I/L	S/L	R
Nooitgedacht Tillite (28°36' S, 24°36' E)				
Total Sample (104)				
$\bar{x}$	6.6	73.4	47.4	15.5
σ	3.2	12.9	14.1	10.3
CV%	49.0	17.7	29.7	66.5
Andesite Clasts Only (88)				
$\bar{x}$	6.9	73.2	47.6	15.3
σ	3.4	13.6	13.6	9.8
CV%	48.9	18.5	28.5	63.8
Cyphex Tillite (29°15' S, 23°24' E)				
Total Sample (94)				
$\bar{x}$	5.9	68.8	45.9	19.3
σ	3.2	12.1	12.4	13.6
CV%	54.2	17.6	27.1	70.7
Andesite Clasts Only (72)				
$\bar{x}$	5.5	69.3	45.6	20.0
σ	2.3	12.2	12.0	12.9
CV%	42.7	17.5	26.3	64.3
Duikersvlei Dropstone Bed (29°15' S, 23°47' E)				
Total Sample (63)				
$\bar{x}$	6.2	73.6	47.0	20.8
σ	6.9	12.7	14.5	9.5
CV%	109.8	17.3	30.8	45.7
Andesite Clasts Only (47)				
$\bar{x}$	5.4	74.9	46.4	21.3
σ	3.0	13.3	14.6	10.6
CV%	55.9	17.8	31.6	49.8

TABLE 4-1 (continued)

	L	I/L	S/L	R
Holpan I, Alluvial Gravels				
Total Sample (117)				
$\bar{x}$	5.7	71.1	37.3	30.8
σ	3.2	15.3	14.0	12.8
CV%	56.0	21.5	37.5	41.6
Andesite Clasts Only (82)				
$\bar{x}$	6.4	71.0	36.9	30.7
σ	3.5	16.2	15.6	13.9
CV%	55.2	22.8	42.4	45.4
Winters Rush I, Basal Gravels[b]				
Total Sample (125)				
$\bar{x}$	5.1	67.8	35.4	35.4
σ	4.0	13.1	13.9	17.7
CV%	78.8	19.3	39.3	50.2
Andesite Clasts Only (64)				
$\bar{x}$	5.2	67.5	37.7	37.2
σ	3.4	13.3	13.8	19.0
CV%	66.6	19.6	36.5	51.1

[a]L is the length (cm) of the longest axis of the clast. I is the length of the intermediate axis, S the length of the smallest axis, R roundness, defined as the percentage of the clast surface which is convex. $\bar{x}$ is the mean, σ the standard deviation. CV% (coefficient of variation) is $100 \cdot \sigma \cdot \bar{x}^{-1}$. The number in parentheses is the sample size. For further information on the sample method and analysis procedure see Appendix B.

[b]See Chapter Five for location and further description.

earlier investigators, form only a minor proportion of the gravels. Often these lithologies are not present at all, and if present, never comprise more than ten per cent of the sample (by individual). This low proportion of dolerite and indurated shale is expectable, given the non-durability of the dolerite to chemical weathering processes, and the limited spatial extent of the indurated shale deposits. Fossil wood clasts from Karoo rocks also occur in the gravels. They may have been introduced from the upper Vaal drainage, or they may have been derived from Ecca shales once found above the Dwyka sediments of the lower basin.

The most important secondary source of clasts is erosion of older gravel beds nearby. There is little doubt that clasts in the Younger Gravels at Windsorton were derived from Wedburg Terrace gravels, that in turn may have been derived from Proksch Koppie gravels, that in turn may have been derived from Holpan gravels.

Consequently, most of the clasts in the alluvial deposits were probably introduced to the Lower Vaal river system through denudation of adjoining hillslopes, and not necessarily as gravel surges introduced from the upstream drainage or derived from active dissection of the pre-Karoo bedrock.

Dating the Older Gravels

Older Gravels features can be arranged in altimetrically descending sequences for several Vaal reaches to produce relative ages (see Fig. 4-20). At Windsorton the Proksch Koppie gravels are clearly older than the Wedburg gravels, and the Holpan gravels are oldest of all. Similarly the Droogeveld gravels are older than the Wedburg Terrace across from Longlands. However, correlating Older Gravels above the Wedburg Terrace in even adjacent reaches is speculative. These features are too limited in extent and spatially

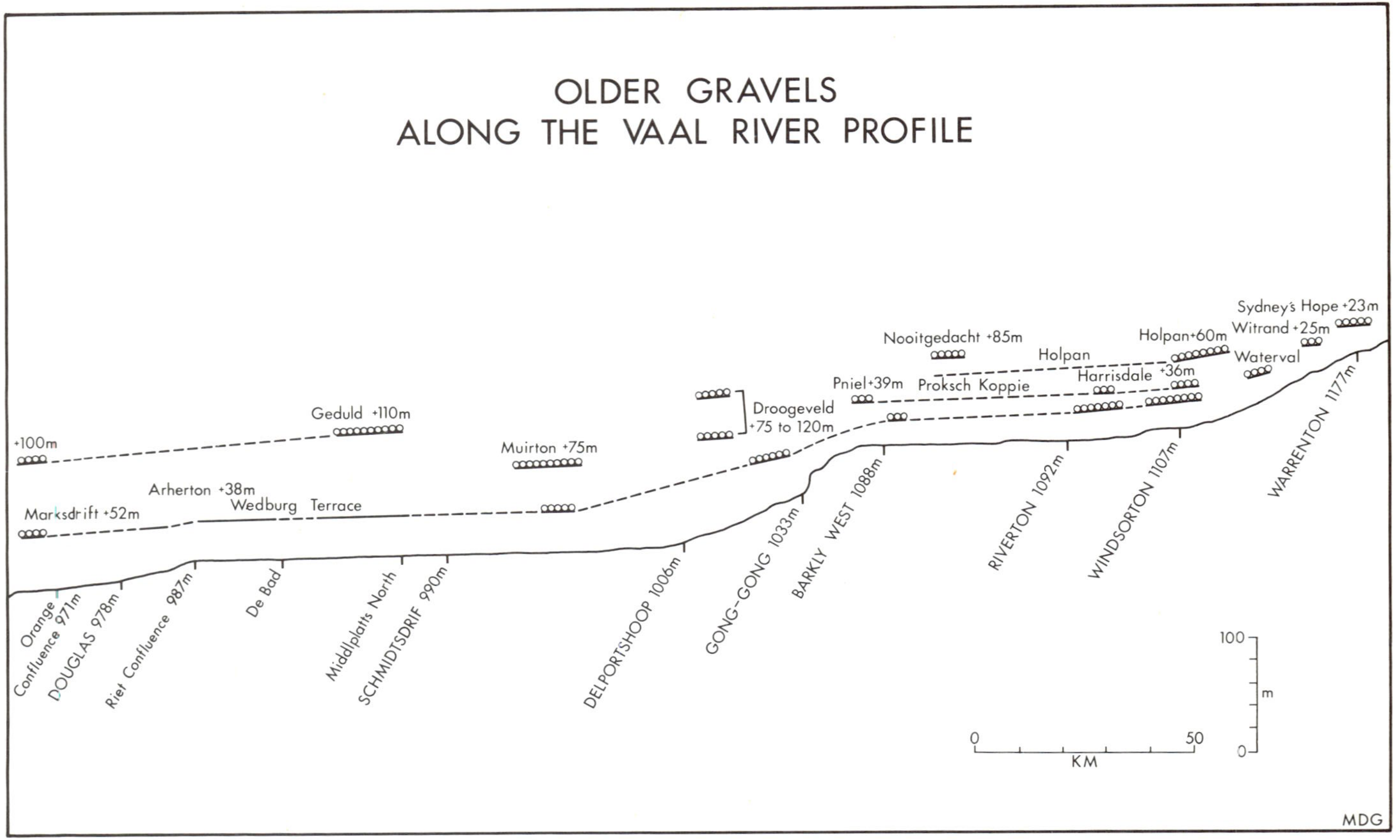

Fig. 4-20. Older Gravels outcrops along the profile of the lower Vaal River.

discontinuous to allow meaningful reconstructions of longitudinal gradients, particularly across the uneven terrains of the Pre-Karoo Surface.

No materials appear to be available for radiometric dating and no fossils have been explicitly reported from the Older Gravels (Helgren, 1977). However, ancient forms of uncertain provenance have been retrieved together with the extensive Younger Gravels fauna (discussed in Chap. 10). These include remains of a giraffoid (*Sivatherium cf. maurusium*; Harris 1976), a suid (*Notochoerus capensis*), a gomphothere, and an assortment of poorly preserved Elephantidae (Cooke, 1949; 1963; 1967; Wells, 1964). *Notochoerus capensis* is also known from the lower Shungura and Mursi Formations of the Lower Omo River Basin and from Kubi Algi near Lake Turkana (Rudolf); a span of 4.2 - 3.2 m.y. B.P. is suggested (Cooke and Maglio, 1972). Maglio's revision (1973) of the early Elephantidae identifies the ancient Vaal elephants as *Mammuthus subplanifrons*, and four fragmentary specimens that cannot be categorized. *Mammuthus subplanifrons* dates from about 4.3 - 2.5 m.y. B.P. in the East African chronologies (Cooke and Maglio, 1972; Coppens, *et al.*, 1976). The *nomina dubia* specimens are named "*Archidiskodon vanalpheni*", "*A. milleti*", "*A. loxodontoides*", and "*Loxodonta griqua*" (Dart, 1929). Maglio (1973) suggests they may represent the same species and be conspecific with *Loxodonta adaurora* or *Mammuthus africanavus*. In either case a date of about 4.5 - 2.5 m.y. B.P. is indicated. Unfortunately the taxonomic position of the gomphothere is unknown, while the sivathere apparently records a species which persisted from the later Pliocene to the mid-Pleistocene (Harris, 1976).

While the exact sources of these fossils are either unknown or poorly recorded, some reasonable speculations can be entertained. The ancient forms and their recorded sites are listed on Table 4-2.

TABLE 4-2

FOSSILS PROBABLY DERIVED FROM OLDER GRAVELS

Specimen	Location[a]
1. *Notochoerus capensis* (left M_3)	Longlands, site unspecified
2. *Sivatherium cf. maurusium* (molar, MMK 3685)[b]	?Gong-Gong or Waldeck's Plant?
3. *Gomphotherium* sp. (molar of Beck, 1906)	?Gong-Gong or Waldeck's Plant?
4. *"Loxodonta griqua"* (molar, MMK 3686)	?Gong-Gong or Waldeck's Plant"
5. *"Archidiskodon vanalpheni"* (left M^3, MMK 4086)	Sydney, "50'-60' depth in 60'-80' terrace"
"Archidiskodon milletti" (left M^3, MMK 4085)	
"Archidiskodon loxodontoides" (left M^3, MMK 4087)	
6. *Mammuthus subplanifrons* (M_3, MMK 3920)	Sydney, "at a depth of 50'-60' in 60'-80' terrace gravels"
7. *Mammuthus subplanifrons* (left and right M_3, MMK 433-4)	Gong-Gong "at a depth of 56' in the middle terrace"
8. *Mammuthus subplanifrons* (molar, MMK 435)	Gong-Gong "deep diggings at a depth of 80'"

[a]Location recorded in the McGregor Museum registry and by Cooke (1949).

[b]MMK numbers denote registration in the McGregor Museum accessions list.

Notochoerus capensis (the type specimen) comes from Longlands, where there are no Older Gravels. Until the 1960's the bar and hotel at Longlands served as the social center for the diggings between Gong-Gong and Sydney, and it seems equally likely that this specimen could have been delivered to the bar at Longlands as having come from any deposits in the Longlands area. Items 2, 3, and 4 on Table 4-2 came from the same collection at about the turn of the century. Dart (1929) and Cooke (1949: 54) attribute them to a Mr. A. Grumpelt, the apothecary at Barkly West, and reasonably suggest these specimens came from Younger Gravels at Gong-Gong or Waldeck's Plant. The other specimens all have better-known provenances presumably coming from the Younger Gravels between Barkly West and Sydney.

As noted in Chapter Three, the Vaal follows a pre-Karoo channelway from Barkly West to Sydney. The Younger Gravels in this sector constitute a thick valley fill, created when Older Gravels deposits and valley-side tillites were eroded and fluvially resorted on the valley bottom. Sporadic veneers of colluvial Older Gravels still are preserved higher up on the valley sides, as is the Wedburg Terrace along the south margin of the valley. The fossils themselves, limited to particularly resistant body parts, are rolled and battered, strengthening the argument for derivation from Older Gravels that no doubt were much more extensive in this sector. The fossils could have come from several of the Older Gravels deposits, but almost certainly not from deposits older than those at Droogeveld. Since they would have been abraded and potentially destroyed each time a new Older Gravels channel was established, it is most reasonable to associate them with the Older Gravels unit that immediately precedes the Younger Gravels, that is the Wedburg Terrace. Consequently, these 4.3 - 3.2 million year old fossils provide the best possible date for the Wedburg Terrace. The other

Older Gravels must be still more ancient, perhaps dating back to the mid-Tertiary, while the capping veneers at some Older Gravels sites may be as recent as late Pleistocene.

Significance of the Older Gravels

The Older Gravels deposits and planation features provide a fragmentary, but valuable record of Cenozoic denudation which has no known parallel in southern Africa. Most Older Gravel vestiges, like the colluvial deposit at Droogeveld, the surficial gravels veneers on the Vaal-Harts interfluve, and the calcreted shale surfaces near De Bad are relatively uninformative, but do record intermittent, coarse gravel transport, local base-level positions, and/or semi-arid, calcrete-producing, rock-weathering regimes. The informative exposures at Holpan and Proksch Koppie, provide details of stream dynamics and suggest environmental fluctuations both somewhat more arid and more humid than present. However, the Older Gravels features have restricted spatial extent and uncertain temporal status. This situation leaves several possible explanations for their development.

Van Riet Lowe (in Breuil, et al., 1948; and 1952)[19] favored a complex explanation which interpreted the progressively lower, Older Gravels features as the result of major climatic fluctuations and periodic breaching of resistant bedrock "barriers" along the Vaal channel. Van Riet Lowe's climatic interpretations matched the then-popular East African pluvial theory, and he correlated gravel aggradation with moderately humid intervals and valley deepening with both arid and very humid intervals. Of course, this processual explanation is as fanciful as are the climatic changes of the East

---

[19]At one time van Riet Lowe (in Breuil, et al., 1948) even speculated on the existence of fault displacement to explain the different Older Gravel levels near Windsorton!

African pluvial chronology.[20] Channel incision was also thought to occur when andesite bars crossing the Vaal were breached by erosion along the channelway. However, as argued above (Chapter Three) the shapes of the modern Vaal valley in the lower basin were effectively formed in pre-Karoo times, and the Dwyka sediments present during Cenozoic exhumation hardly constituted a barrier to stream incision.

In 1947 Cooke argued that the progressively lower Older Gravel deposits were largely related to two postulated river captures. The lowest capture supposedly was at Droogeveld, where a Harts tributary, presumably running from Gong-Gong to Delportshoop, captured the Vaal. As noted above, the deposits at Droogeveld do not support this conclusion, as they are colluvial veneers spread at progressively lower elevations, suggesting gradual movement of the Vaal down the pre-Karoo slope.[21] The other capture occurred upstream, as represented by the gravel "run" (the "London Run") from Schweizer-Reneke to Bloemhof. This run is up to 2 km wide and consists of colluvially reworked, gravel veneers up to 2 m deep, that extend for nearly 50 km. These gravels link the Vaal and upper Harts drainage, but are not paralleled by a modern stream. Here a capture of a Vaal tributary by the upper Harts or even Molopo drainage seems demonstrated, which no doubt had erosional

---

[20]Correlating gravel deposits with humid climates is a repetative theme in the lore of African archeology, a practice seldom questioned by the contemporary earth scientists.

[21]Of course, the colluvial deposits at Droogeveld also do not <u>prove</u> that this river capture did not occur. The Droogeveld gravels could possibly be derived from a single Vaal channelway, but wide extent of the deposits argues that gradual channel migration is the most economical explanation.

and depositional effects in the lower basin during Older Gravels times.[22] However, it is worth noting that the modern Vaal River at Bloemhof has over 60 times the average annual flow of the modern Harts at Taung (see Table 2-5), and the importance of this capture for the Windsorton deposits is problematic.

In 1967 Partridge and Brink formally added two more speculations -- (1) "the headward advance of nickpoints related to subphases of the Post-African erosion cycle" (p. 35) and (2) "increases in the gradient of the river profile through axial warping" (p. 35). The first of these explanations follows from L. C. King's (1963, 1968) interpretations of South African geomorphology where rock-cut terraces can be left in the wake of a presumed retreating knickpoint. As described in Chapter 9 below, King's cyclic knickpoints theory must be rejected principally because their is no evidence for sequential continental uplift of southern Africa or knickpoint retreat in the Lower Vaal Basin. The second notion that the progressively lower, Older Gravels deposits are a result of episodic, tectonic deformation of the upper Vaal basin was first indirectly proposed by Williams in 1932, who traced the discontinuous gravel runs that cross the modern Vaal-Limpopo watershed. Clearly the Vaal had a larger drainage basin during part of the Older Gravels time span. An axis of positive, crustal flexure along the modern Vaal-Limpopo watershed was first formally proposed by A. L. du Toit (1933) and named the "Transvaal-Griqualand Axis". Du Toit pointed out that this line should "...be regarded not so much as an axis of upheaval, but as primarily to the <u>sinking</u> of the ground on its northern side, the faint slope of the southern side being an original

---

[22]DeBeers Consolidated Mines (unpublished) has verified lithologies from eastern Botswana in the "London Run", clearly demonstrating north to south drainage.

feature, and having remained almost untilted" (p. 9, emphasis du Toit's). Despite du Toit's emphasis on the lack of tilting on the south side of this "axis" the arguments have recently gone full circle, with Mayer (1973) concluding that the Older Gravels in the lower basin demonstrate the tectonic deformation of the upper.[23] No doubt the larger, upper drainage basin was important for Older Gravels erosion and deposition, but the role of crustal deformation remains enigmatic.[24]

How, then, are the Older Gravels along the lower Vaal River to be interpreted? First, it must be appreciated that there basically are two categories of Older Gravels -- those of the Wedburg Terrace and those above the Wedburg Terrace. Those above the Wedburg Terrace are predominantly colluvial gravels veneered across generally planar segments of the Pre-Karoo Surface. The few deep deposits with primary alluvial structures are sited in glacially shaped depressions in eroded joint traces. These sequentially lower, Older Gravels are located on progressively lower pre-Karoo slopes,

---

[23]The correlation of rudaceous deposits with positive tectonic dynamism is one of the great sedimentological myths of geomorphology, and cannot be supported (see Garner, 1974: 594-596). The clastic loads of streams are provided largely by erosion of the adjacent slopes. Clast size represents the interaction of rock weathering and erosion processes on local lithologies, which have no systematic relationship to regional tilting.

[24]The nature of the "Transvaal-Griqualand Axis" remains to be studied and mapped in detail. Evidence for deformation consists of apparent warps in planar surfaces along the modern Vaal-Limpopo watershed, and surficial gravel accumulations and other abandoned channelways that connect across the watershed. However, these features can be explained without introducing crustal deformation. The undulations in the planar surfaces may reflect lithological variations or alterations in joint spacing. Or they may be remnants of the Pre-Karoo Surface, that also has not been reconstructed for

so that cyclic denudation during the Cenozoic is not explicitly demonstrated. It seems best to view these higher Older Gravels as an erratic, infrequently preserved depositional residue of a long period of continuing erosion. However, this is not to say that climatic changes did not proceed. Most explanations for the Quaternary glacial episodes invoke "extraterrestial" variations in the earth-sun relationship such as the procession of aphelion and perihelion through the seasons, variations in the solar "constant", and possible clouds of cosmic dust. These sources of global climatic change were as equiprobable in Tertiary as the Quaternary. The changing trends in rock weathering and hillslope erosion, recorded at Holpan and Proksch Koppie probably provide verification of low-order climatic oscillations in the Tertiary of southern Africa.

The Wedburg Terrace does appear to signify an episode of valley incision, followed by an interval of stability in channel levels with little net deposition as lateral planation proceeded to carve the valley shoulders. The subsequent phase of renewed valley incision begins the Younger Gravels record. There are four possible explanations for this behavior: (a) climatic changes, (b) tectonic deformation along the Orange River below the Vaal confluence, (c) tectonic deformation in the upper Vaal basin (through consequent stream captures and/or through an increased Vaal gradient in the upper basin), and (d) river capture in the sense of Cooke (1947).

---

this region. The discontinuous gravel runs could represent the courses of rivers flowing on Karoo-age rocks, above the Pre-Karoo Surface. All of the present deposits appear to be colluvial veneers that no longer preserve primary alluvial structures. Thick deposits appear restricted largely to karstic solution features postdating initial alluvial aggradations (see du Toit, 1951). Mayer's (1973) region-wide speculations on crustal deformation are not supported by field data in the Lower Vaal Basin.

All of these explanations are viable, with some qualifications. Captures by the Limpopo and Natal streams no doubt reduced the upper Vaal basin during Older Gravels times. However, dating and correlating these events to any features in the Lower Vaal Basin is improbable and perhaps impossible. Similarly tectonic deformation, either above or below the Lower Vaal Basin, may have led to cyclic valley incision. While these events are speculative and their investigation essentially unstudied, it is relevant that tectonic displacements below the Vaal-Orange confluence would probably have had negligible impact above the pre-Karoo gorge of the Orange, below Douglas, or above the gorge of the Vaal, below Barkly West. Similarly, the upper Vaal does not run across the presumed axis of uplift, but parallel to it; such deformation would only cause consequent stream captures rather than steepen the Vaal trunk channel.

Regional climatic changes may also have caused the valley incision before and after formation of the Wedburg Terrace. Downcutting could be expected if the Vaal had such large discharges that it was continuously "underloaded" with the excess energy directed toward deepening and widening the channel. Similarly a Vaal with a relative balance between discharge and sediment supply could be expected to maintain near equal erosion and deposition, so that base levels would remain relatively stable.[25] A first clue to the climatic dynamism of the Wedburg Terrace stage is provided by cryoclastic breccias beneath the tufa carapaces along the Gaap Escarpment. These deposits are a broad, morpho-stratigraphic parallel to the Wedburg Terrace, and indicate a decidedly colder regional climate (Butzer, et al., 1978). In this temporal span

---

[25]A full discussion of climatic changes and river incision/aggradation is presented in Chapter 10 below.

cold phenomena from the Southern Hemisphere are now well known. For example, there were extensive piedmont glaciers on the Patagonian Plateau ca. 3.5 m.y. B.P. (Mercer, 1976), and the Antarctic ice sheet was more extensive than today ca. 4.7 - 4.3 m.y. B.P. (Shackleton and Kennett, 1975). A second morpho-stratigraphic parallel to the Wedburg Terrace is the Virginia fossil locality in the Orange Free State, where a comparable fluvial terrace stands above the Sand River (an upper Vaal tributary, see Butzer, 1973). The alluvial deposit yielded a nearly complete Mammuthus subplanifrons, and records a paleo-Sand River with a width and competence far in excess of the modern stream. While the exact relationships of these climatic parallels with valley incision and development of the Wedburg Terrace remains unclear, the theme of Pliocene climatic changes as preclude the major fluctuations of the Pleistocene provides the most viable explanatory vehicle.

Better interpretations of the Older Gravels will only be possible when the other ancient surficial deposits of the South African interior are studied in detail.

CHAPTER 5

THE YOUNGER GRAVELS OF THE VAAL RIVER

> The ancient gravels are found to be extremely rich in implements and faunal remains, and the occurrence as a whole, usually interesting and embracing.
>
> -- C. van Riet Lowe (1935)

Introduction

As the Vaal abandoned the Wedburg Terrace, incision of the Dwyka sediments all along its channel began again. Near Windsorton and between Barkly West and Delportshoop this erosion locally was in excess of 25 m and reached the base of the coincident, pre-Karoo channel. But in other areas, particularly downstream of Schmidtsdrif, incision ceased while Dwyka sediments were still present beneath the channel bed. With this valley exposed, Younger Gravels sedimentation began. At Windsorton the Younger Gravels comprise a complex valley fill, which is overlain by the Riverton Formation. The Younger Gravels sequence at Windsorton is closely replicated below Barkly West at Gong-Gong and Winter's Rush. However, the Younger Gravels are differentiated into two distinct terraces at Winter's Rush that continue downstream to beyond the Orange confluence.

Thick deposits of Younger Gravels that remain today are still actively mined for diamonds. Yields are apparently less than 1 carat per 30 tons,[1] with diamonds localized within fluvially sorted, heavy mineral concentrations. The diggers pits range in size from the traditional 15 x 15 yard "claim" to larger units where eight or more "claims" have been combined. Nearly all the diggers "wildcat", moving erratically and rapidly from one site to another, choosing sites largely on intuition. This mining method leaves cratered landscapes, with the former claims in various stages of collapse as the walls slump and the pits fill with vegetation (Figs. 5-1, 5-2). While more than half of the original Younger Gravels deposits have now been mined, the abandoned diggers pits still provide extraordinary exposures. However, it is an understatement to say that the stratigraphy in these exposures is complex. Individual beds are often capped by steeply dipping unconformities and facies may change rapidly, so that very different exposures may be found in adjacent pits. The present study is based on detailed facies studies and tracing of unconformities leading to the establishment of the type sites and composite profiles described below.

Past investigators of the Younger Gravels have only considered deposits above Delportshoop, primarily those at Windsorton. Despite the many commentaries on the Younger Gravels in both primary and secondary sources, the morpho- and litho-stratigraphies of the Younger Gravels had been investigated only once. As noted in Chapter 1 above C. van Riet Lowe of the Union's Archeological Survey initiated an examination of these deposits in 1936 with P.G. Söhnge and D.J.H. Visser, who had been seconded from the Geological Survey. Their results appeared in a Memoir of the Geological

---

[1]This figure may not be accurate. The diggers tend to be scrupulous about production figures, but keep only casual records of excavation volumes. This sort of data is also of variable utility; size and quality characteristics of the individual stones ultimately determine profits.

Fig. 5-1. Collapsing digger's pit on Riverview Estates. Dwyka shale is exposed at the excavation base, but intact deposits remain in only right background. The largest clasts have 30 to 40 cm long-axes.

Fig. 5-2. Typical, small-scale digger's rig on Riverview Estates. No intact deposits are visible in the photo.

Survey (Söhnge, et al., 1937), however they were less than successful in reconciling their stratigraphies (in particular, see Appendix B of the 1937 monograph, pp. 127-129). The geologists looked at van Riet Lowe's artifact stratigraphy for collaborating evidence, and the archeologist did the reverse. The conclusion was a terrace sequence, subsequently championed by van Riet Lowe (see van Riet Lowe, 1952). He believed that the Younger Gravels near Windsorton consisted of three separate gravel aggradations -- "Gravels I", "Gravels II", and "Gravels III." In addition "Gravels II" was subdivided into "IIA" and "IIB" by a discontinuous "parting". "Gravels I" and "Gravels II" were both part of the +40 ft. terrace, while "Gravels III" were found in the bed of the modern river. Van Riet Lowe further supposed that each gravel deposit rested on a separate bedrock bench, and that each gravel aggradation was succeeded by a deposit of sands -- "Current Bedded Sands", "Calcified Sands", and "Red Sands." Furthermore, each depositional and erosional interval signified a different stage of cultural development as indicated by his typology of the included artifacts. Downstream deposits were then correlated to the presumed events at Windsorton, either on the basis of similar gravel facies or of similarities in the included artifacts.

Not surprisingly this chronology did not prove durable. Van Riet Lowe's categories of handaxes find little theoretical support when compared with modern typologies that emphasize complete assemblages, and they have subsequently been neglected (see Cole, 1961; Mason, 1962, 1967). Similarly the alluvial stratigraphy has not survived persistent criticism (see review in Butzer, Helgren, et al., 1973). As early as 1947 H.B.S. Cooke pointed out that "Gravels I" occurred only at Windsorton, and that "Gravels II" and "III" appeared to merge locally. Cooke speculated that the gravels formed a conformable set, succeeded by the "Calcified Sands." Partridge and Brink (1967: 25) offered a more conservative interpretation, lumping the Younger Gravels and associated sands at Windsorton into one unit, called the "Current Gravels", referable to "a single cycle of channel development. . . culminating in the development of the modern floodplain."

The results of this investigation are substantially contrary to all the earlier reports. In particular, the irregular bedrock channel of the Vaal at Windsorton (as at Barkly West, Gong-Gong and Winter's Rush) was cut to modern dimensions prior to and during the first aggradation of the Younger Gravels. The bedrock benches envisioned at Windsorton by van Riet Lowe are not present. The Younger Gravels are composed of three distinct, depositional units--Younger Gravels A, B, and C--here designated formally as the Rietputs Formation, Members A, B, and C.[2] Episodes of erosion and pedogenesis intervene between these units. This depositional framework is well preserved at Windsorton, Gong-Gong and Winter's Rush while further downstream the Younger Gravels record consists of separate terraces with shallow depositional veneers. Following the Younger Gravels Rietputs Formation is the Riverton Formation, a complex, five-fold, cut-and-fill sequence of sands and silts defined in five members (designated from oldest to youngest--I, II, III, IV, and V; Butzer, Helgren, et al., 1973). Consequently a total of

---

[2]Rietputs A, B, C and Younger Gravels A, B, C are equivalent respectively. This nomenclature has been revised from that given in the preliminary report by Butzer, Helgren, et al. (1973), since gravels B and C of the 1973 report have proven to be conformable and are here designated Gravel B (Rietputs B). The Gravel C (Rietputs C) referred to here is found in or near the modern river, and is substantially similar to the "Gravels III" of van Riet Lowe. In principle, the term "Younger Gravels" should no longer be used, both because it describes a lithofacies and because it has been used uncritically to categorize ancient alluvia elsewhere in southern Africa. However, the term "Younger Gravels" has long been popular, and its informal use is continued for the sake of convenience.

eight depositional cycles and at least as many paleosols are compressed into the alluvia along some reaches of the modern Vaal. Furthermore the alluvial fills may be marked by as many as five distinct terraces. Van Riet Lowe's (1952) "Gravels I" and "Gravels IIB" at Windsorton are partly referable to our Gravel B (Rietputs B) while his "Gravels IIa" at Windsorton and "Gravels II" downstream of Windsorton broadly correspond to our Gravel A (Rietputs A). Van Riet Lowe's sand units variously refer to sandy facies of our Gravel B (Rietputs B), as well as to Members I, II, III, and IV of the Riverton Formation.

The emphasis of this investigation was explicitly directed towards the Older and Younger Gravels. The Riverton Formation was studied at key Younger Gravels localities, with most concern placed on the contacts as well as on the somewhat enigmatic, lowermost members (I and II). The Riverton Formation in the Lower Vaal Basin is mapped here, but not all of the key Riverton profiles are described here, e.g. the holotype at Riverton on Vaal (Butzer, Helgren, et al., 1973) or the equally important sections along the lower Riet River (Butzer, in preparation). Since the Riverton Formation

---

The Rietputs Formation takes its name from the original farm "Riet Puts" (meaning "reed well") which occupied the east side of the Vaal at Windsorton. The sections described here at Riverview Estates and in the modern Vaal channel serve as the stratotype, with the deposits described below at Winter's Rush as a reference stratotype. The upper boundary is the Riverton Formation. The origin and paleontology are described in Chapter 10 below. The spatial extent is along the Vaal channel from Windsorton to the Orange River confluence. The similar deposits along the Riet, Harts, and Orange Rivers are informally listed as equivalents to Rietputs Members A, B, and C. No formal morpho-stratigraphic nomenclature is proposed here, largely because neither the U.S. nor South African stratigraphic codes adequately negotiates the combined terrace morphologies and lithofacies described here. Informally defined are the terraces of Rietputs Formation Member A and B; the equivalent terraces on the Harts, Riet, and Orange rivers are informally designated Rietputs A and Rietputs B.

will be outlined in further detail at a later time, the depositional patterns, relict pedogenesis, and inter-member unconformities of Members III, IV and V (which comprise the modern floodplain surfaces) are only briefly noted in this study. The Riverton Formation provides classic examples of fine-grained alluviation during a period of fluctuating regional climate. In addition, abundant archeological materials remain to be recovered by systematic excavation.

This chapter describes the litho- and morpho-stratigraphy of the Younger Gravels--Rietputs Formation--at each important site along the lower Vaal River.

The Rietputs Formation at Windsorton

The classic occurrence of the Younger Gravels is at Windsorton, where the Rietputs Formation and earlier Riverton Formation are part of one +12 to 14 meter floodplain terrace (the "+40 ft terrace" of Söhnge and Visser, 1937; Figs. 5-3, 5-4, 4-17). Although this surface has been reached by historic floods, there is only a thin veneer (up to 20 cm) of recent sediments. A lower, +8 to 9 m terrace (the "+25 ft terrace" of Söhnge and Visser, 1937) also occurs along both sides of the modern channel at Windsorton. This terrace is composed of grayish-brown sandy silts pertaining to Member IV of the Riverton Formation; also present at this level are minor erosional shoulders cut into the higher terrace. Prior to the diamond rush, a +4 to 5 m terrace may also have been present at Windsorton. Remnants of the grayish-brown, sandy silts of Member V of the Riverton Formation are found at this level, but construction of coffer dams ("breakwaters") in the modern river by diggers and mining on the channel fringes has nearly obliterated the 19th century channel morphology.

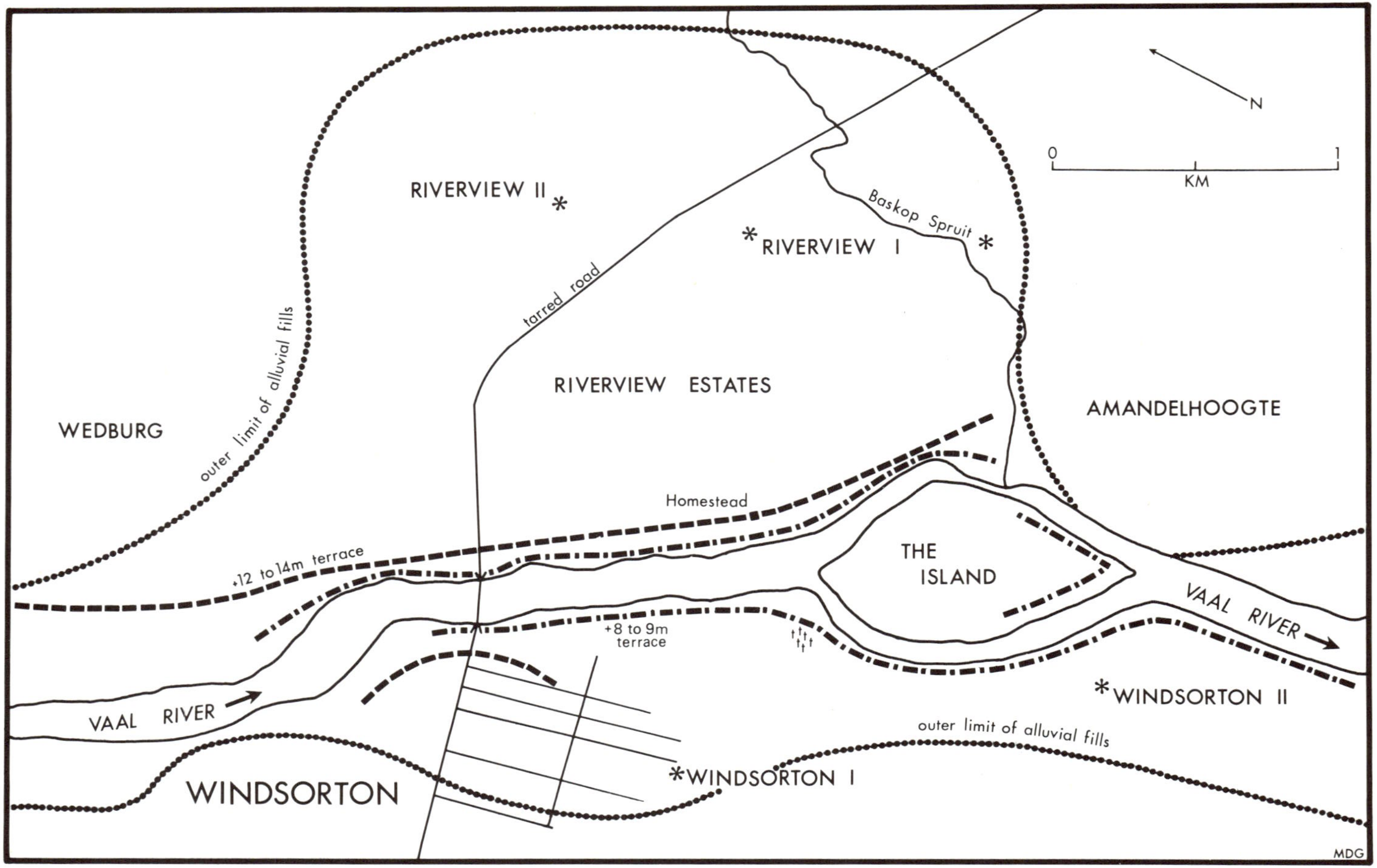

Fig. 5-3 Sedimentary localities near Windsorton

Fig. 5-4. Vaal channel at Riverview Estates. Homestead is at right center. In the right foreground is the north end of The Island with abandoned diggings in the channel.

The gravels near Windsorton have been strongly emphasized in earlier reports due to their artifact and diamond content. However a nearly equal volume of arenaceous sediments is also present. Thick gravel deposits occur east of the Vaal to the north and east of the Riverview Estates farm buildings and at the lower level on Wedburg (van Riet Lowe's "Gravels IIA and IIB"), and west of the Vaal below the Windsorton "main street" (van Riet Lowe's "Gravels I"). Thinner gravel fills are found in the Vaal channel and beneath a thick layer of Riverton Formation sediments, both on the southern part of the Windsorton townlands and on the peripheries of the thick gravels on Riverview Estates.

Unlike the usual Older Gravels outcrop, the gravels of the Rietputs Formation near Windsorton are often thick and record primary depositional structures with significant facies variations. Söhnge and Visser (1937: 26-33) describe the rapid thickness changes of the sand and gravel layers between many individual pits and outline the irregular occurrence and close juxtaposition of the andesite, shale, and tillite bedrocks.[3] Some of the small pits reported by Söhnge and Visser are still partly exposed, but most have been refilled. However, diamond digging has continued, using ever larger equipment, so that many new pits are available. Several excavations are quite large, and some are located in areas unexposed to earlier investigators.

The Rietputs and Riverton formations at Windsorton can be best understood in terms of five key exposures: Riverview I, Riverview II, Breakwater, Windsorton I, and Windsorton II. These sites provide a depositional, erosional, and pedogenetic framework within which all the remaining deposits can be categorized. The fill to the east of the modern Vaal on Riverview Estates is described first.

Riverview I. Riverview I is located on the +14 m terrace about 900 m east-northeast of the Riverview Estates farm buildings[4] (see Fig. 5-4). This excavation with a diameter of more than 70 meters provides the largest exposure on Riverview Estates. Historically this area had only been mined by selective tunneling (Figs. 5-5, 5-6). Further textural data are in Appendices A and B.

---

[3]A small dolerite sill and a kimberlite fissure also occur on Riverview Estates.

[4]The Riverview Estates farm buildings are the "Homestead" of Söhnge and Visser (1937).

Fig. 5-5. Riverview I exposure. Boundary between Rietputs A and B is particularly well marked.

RIVERVIEW I

Unit 1. (top) "Soil", 0.4 to 0.6 m apedal, weakly stratified, dark reddish brown (5 YR), silty sand (no. 022), re-worked by the early diggers at Windsorton.

-- gradual, wavy, unconformable contact --

Unit 2. "Calcrete", 2.0 to 2.5 m, moderately cemented, white to very pale brown (10 YR), silty sand (no. 031A) with root hollows with up to 3 cm diameters, which contain weakly-cemented, reddish brown (5 YR), silty sand (no. 031B).

-- clear, wavy, unconformable contact --

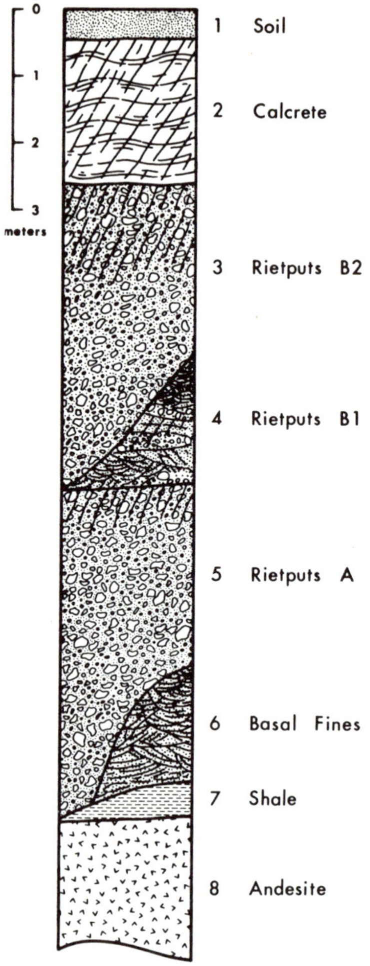

Fig. 5-6. Generalized sedimentary profile at Riverview I.

Unit 3. "Rietputs B2", 2.5 m, crudely horizontally-bedded, granule to coarse cobble conglomerate with a white to very pale brown (10 YR) matrix of very poorly sorted, silty sand. The sub-rounded gravels are arch-typical with coarse pebbles and fine cobbles as the most frequent size-classes and andesite as the most frequent lithology. The matrix at the top of the unit is strongly cemented (no. 030), while at the base it is only weakly cemented (no. 029).

-- gradual to abrupt, wavy, conformable contact --

Unit 4. "Rietputs B1", 0 to 2.0 m, uncemented, complex interbeds of gravels, sands, and silty clays. The gravels vary from the repetitive, poorly sorted, crudely-bedded, granule to coarse cobble conglomerates to well sorted and rounded conglomerates in both the fine cobble and fine pebble-grades, with distinctive bedding planes dipping at up to 31°, suggestive of migration of midchannel bars or dunes. The yellowish brown (10 YR) to pink (7.5 YR) and very pale brown (10 YR) gravel matrices range from medium to coarse, quartzose sands (nos. 025, 028) that include clasts derived both from erosion of previously calcreted gravels and from erosion of an oxidized gravel matrix. The reddish yellow (5 YR, 10 YR) to very pale brown (10 YR) sandy beds which interdigitate with the gravels, vary from medium to fine-grade (nos. 026, 027) with crude, horizontal bedding, and rarer tabular and trough cross-bedding. Clasts of derived calcrete also occur in these sands, as do dispersed pebbles. The light gray (N, 10 YR) to very pale brown (10 YR), silty clays are in part vertisolic and interdigitate with the sands.

-- abrupt, wavy, unconformable contact which gently undulates at less than 1° in all directions --

Unit 5. "Rietputs A", 4.8 to 5.5 m, crudely stratified, granule to boulder-grade conglomerate. The white and pale brown (10 YR) (nos. 024, 022) matrix of the erosionally-truncated top of the unit is strongly cemented (indurated) with secondary and tertiary calcite re-crystallization, but the carbonate enrichment and cementation gradually decreases to zero downward as matrix colors change to light gray (2.5 Y) and finally, oxidized, brownish yellow (10 YR) in the zone of modern groundwater fluctuation (nos. 018, 019, 020, 021, 023). The inhomogenous matrix textures vary from silty sands to fine, medium, and coarse sands. The subangular to well-rounded gravels are arch-typical with coarse cobbles and boulders packed with all classes of finer grade clasts.

-- abrupt, wavy, unconformable contact --

Unit 6. "Basal Fines", up to 2.1 m, uncemented, chaotically-bedded complex of silts, silty sands, and sandy silts, including dispersed cobbles and pebbles, particularly at the base. Colors vary from light gray (5 YR, 5 Y) to yellow (2.5 Y, 10 YR) to white (2.5 Y) and to very pale brown (10 YR) (nos. 010, 011, 012, 013, 014, 015, 016, 017, 097, 098, 099). The bedding is partly fluvial, but primarily the product of bankfall and slumping, with secondary, compressional deformation by the capping gravels units. The silts are frequently quartz flakes or weathered quartz grains, and the sand fractions include clasts of derived, cemented gravels matrix.

-- abrupt, wavy unconformable contact --

Unit 7. "Bedrock", both Dwyka shales and Ventersdorp andesite.

In sum, above the irregularly-exposed, Pre-Karoo Surface at Riverview I there is a suite of fine sediments, two distinct gravel aggradations, and a capping calcrete and soil. The "Basal Fines" should be viewed as a collage of the fine sediments present on the Vaal floodplain and channel margins during the final stages of valley downcutting and during the first stages of aggradation of Member A of the Rietputs Formation. Other fine-grained facies of Rietputs A are not present near Riverview I, nor are such sediments known anywhere near Windsorton. Monotonous rudaceous deposits like Rietputs A (and most primary Older Gravels features) suggest a winnowed lag, with the sandy matrices being "trapped" in the hydrodynamic "shadows" around coarse clasts as the finer fractions were generally flushed downstream.[5] Immediately southwest of Riverview I,

---

[5]Genesis of the relict gravel deposits at the base of many modern streams is a problem that has received little attention. Studies of modern streams with coarse sediment loads (e.g. Williams and Rust, 1969; Rust, 1972) clearly demonstrate that a variety of finer sediment forms are typical. However, during high-magnitude

"Rietputs A" reaches the top of the +14 m terrace and laps against a shale ridge that is also planed at +14 m. In this area and at other isolated sites on Riverview Estates where Rietputs A is at the surface there is a cambic paleosol, which is little more than a rubefied rock weathering zone, but it is nonetheless distinct from the usual calcretes. Colors reach red and dark red (2.5 YR; no. 073), and were produced by selective weathering of ferromagnesian minerals. At depth these exposures are enriched in carbonates, but any calcrete in this setting is a secondary product of later precipitation from throughflow waters. This cambic paleosol antedates all calcretes in the valley fills at Windsorton, and suggests a lengthy interval between deposition of Rietputs A and Rietputs B.[6]

Returning to the Riverview I exposure, the impressive unconformity between Rietputs A and B probably owes its morphology to lateral stream erosion as the Vaal channel was incised following aggradation of Rietputs A. Still later, the surface was cemented, probably by precipitation from throughflow waters draining the shale and dolerite terrains to the east, rather than by evaporation of ponded floodwaters. The calcrete detritus in Rietputs B1 indicates this cementation was widespread on gently undulating terrains below the +14 m terrace, before Rietputs B was deposited. Rietputs B1 and B2 signal renewed large-scale mobilization of coarse clasts near Riverview I. Reddish yellow and yellowish red sands are widespread and were reworked with the finer gravels into complex

---

flows, when coarse clasts are mobilized, fine sediments appear to be flushed away and are not preserved in the ultimate depositional residue.

[6]The stratigraphic position of this cambic paleosol is also supported by the derived, rubefied sands in "Rietputs B1". Perhaps this soil development corresponds with the general rubefaction of the upper horizons at several Older Gravels sites (see Chap. 4).

patterns of dunes and braid-bars with slackwater muds. Eventually the massive gravels of B2 were deposited, and yet another interval of calcretion occurred. The basic sediments of Unit 2 are colluvial/ alluvial sands from beyond the east margin of the floodplain. These sands were subsequently calcreted (again, non-pedogenetically, from evaporation of throughflow waters). The detritus in this calcrete is the local facies of Member III of the Riverton Formation. The "soil" of Unit I is basically a floodplain--margin facies of Member IV of the Riverton Formation, but it has been greatly modified since European settlement.

Riverview II. Riverview II is another, large, recent excavation on the +14 m terrace on Riverview Estates. The pit is 15 x 45 m and located 800 m north of Riverview I, 1400 m northeast of the Riverview Estates farm buildings (see Figs. 5-3, 5-7). Further textural data are in Appendices A and B.

RIVERVIEW II

Unit 1. "Soil", 0.7 m, modern soil zone. The 0.3 to 0.4 m A-horizon is slightly hard, reddish brown (5 YR 4/4), silty sand with coarse, angular blocky structure and dispersed, fine, $CaCO_3$ concretions (no. 033). The 0.3 to 0.4 m B-horizon is moderately hard, reddish brown (5 YR 5/5), silty sand with coarse, angular blocky structure and common, fine, $CaCO_3$ concretions (no. 034).

-- gradual, wavy unconformable contact --

Unit 2. "Calcrete", 3.2 m in total. The top 0.2 to 0.3 m is white (10 YR), cemented, very fine calcrete breccia of coarse angular blocks of clayey silt sand (no. 035). Medially is 1.7 to 1.9 m, strongly cemented, pink (5 YR) to white (10 YR) sandy silt to silty sand, which has reddish yellow (5 YR) and black oxidation stains (nos. 036, 037). At the base is 1.0 to 1.1 m

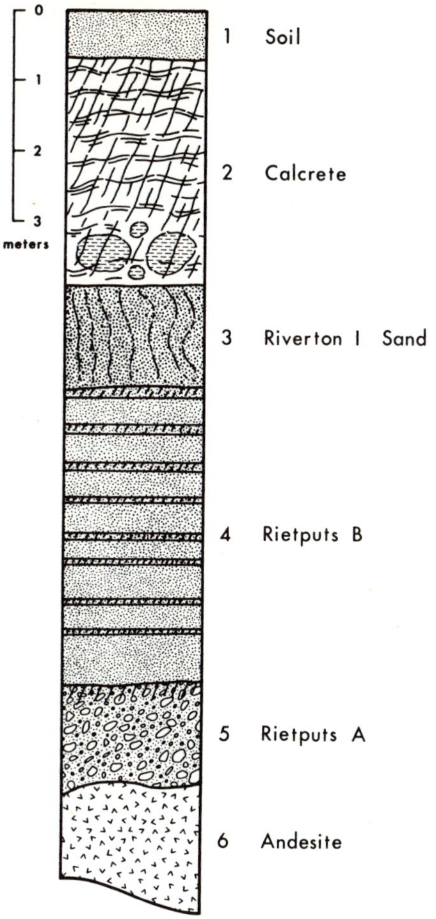

Fig. 5-7. Generalized sedimentary profile of Riverview II.

of light gray (5 YR), moderately cemented silty sand (sample 038), which also has reddish yellow oxidation stains. This lowest zone includes spheroidal inclusions (with diameters up to 0.4 m) of dark brown (10 YR) and brown (7.5 YR), soil-like material, which texturally are silty sands (nos. 039, 040).

-- gradual, smooth, unconformable contact --

Unit 3. "Riverton I Sand", 1.4 to 1.5 m, strongly cemented, reddish yellow (7.5 YR) silty sand with crude horizontal bedding (no. 041).

-- gradual, smooth, unconformable contact --

Unit 4. "Rietputs B", 4.8 m total thickness, cyclic sedimentation set of thin, cemented layers of silty sand and thicker, uncemented layers of silty sand. Calcium carbonate, root casts ("root drip") and root hollows of pene-contemporary vegetation occur throughout. The cemented horizons with colors of white (10 YR), very pale brown (10 YR), gray (N, 7.5 YR, 2.5 Y), and light gray (5 Y) vary in thickness from 0.1 to 0.3 m and exhibit crude, horizontal bedding, and include areas of secondary and tertiary calcite re-crystallization (nos. 042, 044, 046, 048, 050, 052, 053). The pale yellow (2.5 Y) and very pale brown (10 YR), uncemented, horizons have slightly coarser textures, and vary in thickness from 0.4 to 0.7 m and have crude horizontal bedding, and rare, local areas of both planar and trough cross-bedding (nos. 043, 045, 047, 049, 051, 054). Each layer is horizontal and is depositionally unconformable relative to its bounding layers.

-- abrupt, smooth unconformable contact --

Unit 5. "Rietputs A", 3.0 m, typical granule to boulder-grade conglomerate with crude, horizontal bedding. At the top of the unit the silty sand matrix is strongly cemented (no. 055, 056), but the cementation gradually decreases downward. The gravels at the upper boundary of the unit exhibit re-sorting diagnostic of erosional truncation.

-- abrupt, wavy, unconformable contact --

Unit 6. "Bedrock", Ventersdorp andesite and Dwyka tillite.

To sum up, the bedrocks at Riverview II are followed by: a gravel deposit, a cyclic aggradation of cemented and uncemented sands, a separate sand horizon, and a capping calcrete and soil. Based on their comparable depth, position relative to bedrock, cementation, and erosional truncation, Rietputs A at Riverview II can be correlated with Rietputs A at Riverview I and these units presumably share the same depositional history. After Rietputs A was aggraded, this deposit was eroded, probably by a tributary stream cutting to a lower base level, and the upper surface was cemented (calcreted). This cementation is again not pedogenetic, but the result of precipitation from throughflow and surface waters draining the terrains east of Windsorton. Above the gravels is the cyclically-deposited Rietputs B. This sequence of cemented and uncemented layers seems best interpreted as the result of tributary and colluvial sands washed and sorted along a lateral erosional channel or high-flow channelway during a flood, and then secondarily cemented with precipitates from throughflow and surface waters draining the area east of Windsorton. However, the extensive "root-drip" indicates that a low flow-energy or ponded setting was typical most of the time.

Rietputs B at Riverview II generally correlates with Rietputs B at Riverview I, but at Riverview II is 3 m lower in depth. Rietputs B at Riverview I probably represents a gravel braid-bar complex while Rietputs B at Riverview II preserves a more typical channel-margin facies. Strict synchrony cannot be assumed. Above Rietputs B at Riverview II is a tributary-sand facies of Member I of the Riverton Formation. These sands were then also truncated by erosion, and the capping calcrete and, still later, the parent material for the soil was deposited. This calcrete also is the cemented alluvium of a lateral tributary. The inclusions of soil derivatives at the base of this calcrete also suggest complex depositional interaction between the main river and the lateral stream. This unit is referable to Member III of the Riverton Formation, while the soil-like masses are perhaps all that remain of Member II.[7] The modern soil is developed in more recent floodplain sediments, a local variant of Riverton Formation, Member IV.

The exposures at Riverview I and II provide reliable themes for easy categorization of the other exposures on Riverview Estates located more than 100 m from the modern Vaal. West of Riverview II the major change is the appearance of higher-energy textures and structures in Sands B and the Riverton I Sands. Both units locally display high-angle, planar and trough crossbedding. The surface of the Rietputs-B sandy facies at Riverview II undulates significantly, while the smooth, eroded surface of Rietputs A is typically present beneath. General bedrock configurations at Windsorton certainly suggest that the formative channelways should have been located west of Riverview I and II, where Rietputs B1 pinches out and the +14 m terrace is composed of Rietputs B2 over Rietputs A. Expectably the intervening unconformity often is poorly preserved, but it is well marked locally and the heavy mineral concentration at the base of Rietputs-B gravels shows up nicely on the diamond diggers' sorting tables. Rietputs B typically is calcreted, except below the andesite ridge at Wedburg which shielded the deposits from carbonate charged throughflow waters draining the terrains east of Windsorton. East of Riverview I and II, the exposures usually are similar to Riverview II, with the sandy facies of Rietputs B replaced by typical,

---

[7]The east (left) bank "Calcified Sands" and "Current Bedded Sands" noted by van Riet Lowe (1952) generally correspond to the sands and calcrete above the gravels at Riverview II. As reported by van Riet Lowe these arenaceous sediments included Achatina remains. No molluscan materials were found during this investigation, but are found in the top of Member III at the Riverton holotype (Butzer, Helgren, et al., 1973).

coarse gravels, but still capped by Riverton I Sands and a layer of calcreted sands like the "Calcrete" at Riverview I. In addition, thin wedges of clayey and sandy silts (referable to Riverton II) are locally present above the Riverton I sands. There are few exposures south of Riverview I. Most of the terrace surface is occupied by Riverton III and IV sands that now form irrigated fields. At one exposure a thin layer of gravels was present over truncated shales. In addition, beveled shale surfaces are present. The south side of the alluvial fill on Riverview Estates appears to be most analogous to the thick sand deposits at Windsorton (see below); but only fragmentary exposures were available.

An important additional facies of Rietputs B is preserved on the lower level at Wedburg. Here 1 to 2 m, yellowish red (5 YR), inhomogeneous sand (no. 254) rests on Rietputs-A gravels and below Riverton III silty sands. These Rietputs-B colluvial sands contain dispersed pebbles and provide evidence of significant hillslope erosion coinciding with Rietputs-B deposition.

The exposures between Wedburg and the Riverview Estates farm buildings are extraordinarily complex (Figs. 5-8, 5-9); and the terrains have been heavily mined (Figs. 5-1, 5-2). Both the +9 m and +14 m terraces are present and within them facies and unit thicknesses change rapidly. Systematic understanding is only possible through an appreciation that this area recurrently has been the axis of the principal Vaal channelway since Rietputs-A deposition. Consequently coarse-grained sedimentation with high-energy structures is typical and the intervening unconformities often dip steeper than 5°. A thin layer of uncemented Rietputs-A gravels is usually present above bedrock, but calcreted Rietputs-B gravels dominate the base of most exposures. Above these gravels, gravel and/or sand facies of Riverton I (usually not calcreted, 7.5 - 10 YR 5-7/3-6 colors) may be present, as well as gravel and/or sand facies of Riverton III (usually calcreted, 10 YR - 5 Y

Fig. 5-8. Typical Riverview Estates gravel exposure. Here, about 200 m northeast of the Homestead, nearly 2 m of calcreted Rietputs B gravels are surrounded by diggers' debris.

white colors) and rare, fine-textured facies of Riverton II (see Helgren, 1978). At the surface are sand and/or gravel tributary facies of Riverton IV (with 5 - 7 YR reddish yellow and yellowish red colors) and silt and/or gravel facies of Riverton IV (with 10 YR 5-6/3 brown colors). Complicating this picture are at least two shale pedestals that intersect the +14 m surface as well as the carefully sorted residuals of 100 years of diamond digging.

Breakwater. After the complex of sands and gravels designated here as Rietputs B had been aggraded, the Vaal and its tributaries at Windsorton cut down once again, after which most of the gravels lining the modern channel were deposited. Not surprisingly, these gravels are seldom exposed and few occurrences have survived the diamond diggers. Outcrops are preserved only behind the "breakwaters"

Fig. 5-9. Riverton Formation sands and gravels at Riverview Estates; about 100 m north of the Homestead. Riverton IV, red colluvial sands (30 m thick) cap calcreted Riverton III sands, Riverton I sands and gravels and Rietputs-B gravels.

of diamond diggers and rarely on channel fringes. These deposits are Rietputs Formation Member C (Figs. 5-10, 5-11).[8] During this investigation Rietputs-C exposures were available behind breakwaters in the Vaal channel north of the Windsorton townlands and again south of The Island, while channel fringe exposures were available near The Island.

Rietputs-C deposits are extraordinarily similar throughout the basin. Textures are within the range of all the other gravels of the Rietputs Formation (see Slangheuvel gravel analysis in Appendix B), but matrix colors and cementation respond to their peculiar hydrologic situation.[9] In areas below the modern watertable, reduced olive gray (5 Y), green (5 G), and light or dark blue (5 B) are typical matrix colors, while the capping horizon is compacted with ferromanganese and limonitic cements. Where above the watertable, as on channel margins, matrix colors, vary from pale brown (10 YR) to red and reddish yellow (2.5 - 7.5 YR) while the capping horizon is compacted and cemented by carbonates. Transitional settings are known. Total thicknesses vary from 1 to 6 meters. Rietputs C has provided most of fossil mammal collection from the Lower Vaal Basin (Helgren, 1977). None of the sediments observed above Rietputs C at breakwater sites could be assumed independent of the last century of diamond digging. Finer textured facies of Rietputs C or equivalent overbank deposits are not known.

---

[8]During this investigation more than two dozen outcrops of Rietputs C (Younger Gravel C) were discovered and studied. At one time these deposits were apparently continuous between Windsorton and Schmidtsdrif. Contemporary diggers generally believe no primary deposits are left in the modern channelway anywhere in this sector, but deposits do persist, at least locally, beneath Riverton fill.

[9]Accurate identification of gravel units in the Vaal channel is potentially difficult. During this investigation only gravels of Rietputs C and rare gravels of the Riverton Formation were observed.

Fig. 5-10. Vaal channel behind breakwater, south of The Island, 1973. In the foreground modern Vaal sediments cap calcreted gravels of Rietputs C.

Fig. 5-11. Vaal channel south of The Island, with the breakwater in Fig. 5-10 now flooded. In the foreground are irrigated fields on Riverton Formation deposits.

The valley fill on the Windsorton side of the Vaal is dominated by sands and silts of the Riverton Formation. At depth, variable thicknesses of gravels are present over shale, tillite, and andesite bedrocks. These deposits are summarized here in terms of two similar exposures -- Windsorton I and Windsorton II. Below the Wedburg Terrace only one morphologic terrace is present on the Windsorton side of the Vaal, the +8 to 9 m, Riverton IV, which is capped by yellowish red (5 YR) sandy colluvia. The net effect is a gradually rising surface, from the +8 m terrace to +15 m at the edge of the valley fill. The +12 to 14 m terrace is locally preserved beneath this ramp.

Windsorton I is located 0.9 km south of the tarred road. The surface is at +13 m, immediately below the slope leading up to the Proksch Koppie Platform and only 60 m from the margin of the valley fill (Figs. 5-3, 5-12, 5-13).

WINDSORTON I

Unit 1. (top) "Red Sand", 0.3 m, apedal, red (2.5 YR) silty sand (no. 077). This unit is the colluvial facies of early Riverton IV.

-- abrupt, smooth unconformable contact --

Unit 2. "Calcrete", 1.2 m, reddish yellow (5 YR), moderately cemented, silty sand and sand (no. 079). This unit is the capping calcrete on Riverton III.

-- diffuse, smooth, conformable contact --

Unit 3. "Riverton III Sand", 3.1 m, uncemented, very pale brown (10 YR), silty sands with rare, dispersed pebbles and crude horizontal bedding (no. 080). Present are carbonate-cemented "rootdrip", and rootlet fillings and inter-stratified wedges (up to

Fig. 5-12. Windsorton I exposure; gravel level is at very base of photo.

10 cm thick) of reddish yellow (5 YR), slightly coarser, silty sand (no. 081). Many of the sand-sized clasts are weathered, flake-shaped quartz.

-- gradual, smooth unconformable contact with local pebbles --

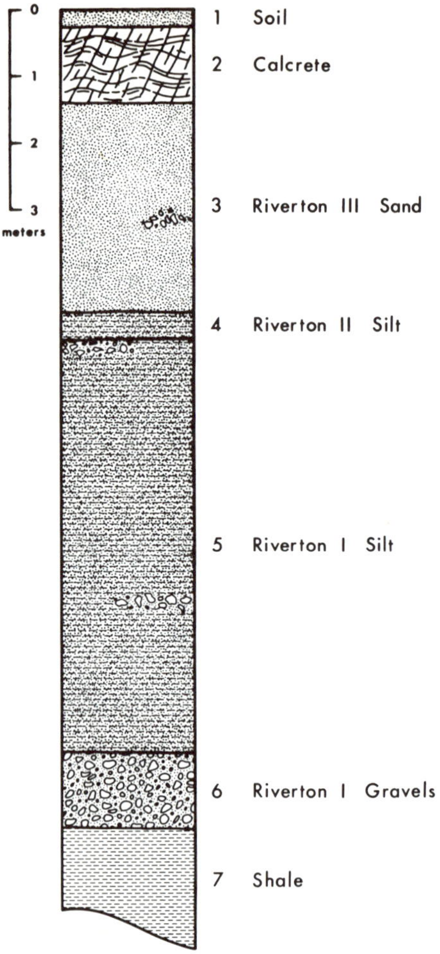

Fig. 5-13. Generalized sedimentary profile at Windsorton I.

Unit 4. "Riverton II Silts", 0.3 m, uncemented, very pale brown (10 YR) sandy silt with weakly developed, horizontal stratification (no. 082).

-- gradual, smooth, unconformable contact with local pebbles --

Unit 5. "Riverton I Silts", 7.0 m total thickness dipping at less than 1° toward the Vaal with weak horizontal stratification and rare, low-angle planar and trough cross-bedding in coarser textured localities. The upper 2.6 m is moderately-cemented, very pale brown

(10 YR), sandy silt (no. 083). The lower 4.4 m is uncemented, very pale brown (10 YR), silt (no. 084), with "rootdrip" and repetitive lenses (up to 30 cm thick) of very pale brown (10 YR), silty sand (no. 085). Many of the sand-sized clasts throughout this unit are weathered, flake-shaped quartz.

-- abrupt, smooth, conformable contact --

Unit 6. "Gravels", 1.5 m typical, granule to boulder-grade conglomerate with an uncemented matrix of very pale brown (10 YR), inhomogenous sands (no. 086). These gravels are referable to Riverton I.

-- abrupt, smooth, unconformable contact --

Unit 7. "Bedrock", Dwyka shale.

In total then the Windsorton I exposure records both channel and channel-margin facies of several elements of the Riverton Formation. Unit 1 is a slope wash of early Member IV, and Members III, II and I are stratified below. The gentle dip of the lower units toward the modern Vaal suggests the coeval channelways were all located to the east, while the weakly developed bedding features suggest a terrain of persistently low flow-energy and colluviation from the adjacent slope. The cementation of units 2 and 5 (Riverton III and I) again records calcretion via precipitation from throughflow waters; it is coeval with the "root drip" in the respective units.[10] In addition, the total thickness of the "fining upward" Riverton Member I is particularly impressive and records a primary eastward channel shift in early Member I times.

---

[10]Calcretion is much less impressive in the valley fill west of the Vaal. This is a reflection of the different settings with respect to passage of ground water and throughflow waters. Riverview Estates is open to the east to drainage from the shale and dolerite terrains that extend beyond the Orange Free State boundary. However the fill on the Windsorton townlands is at the base of the andesite

Windsorton II is located 2.1 km south of the tarred road. The exposure is at +11 m opposite the south end of The Island. It is 500 m from the west margin of the valley fill and only 50 m from the terrace edge (Figs. 5-3, 5-14).

WINDSORTON II

Unit 1. (top) "Red Sand", 1.3 to 1.5 m, uncemented yellowish red (5 YR) silty sand (no. 312) with rare isolated pebbles and weak vertical structuring superimposed on crude, horizontal (largely colluvial) stratification (dipping at less than 1° toward the modern river).

-- abrupt, smooth to wavy, unconformable contact --

Unit 2. "Riverton III Silt", 0.9 to 1.0 variably cemented, light yellowish brown (10 YR) clayey sand silt (nos. 311, 314) with well developed vertical structuring, local (horizontal) stone lines, abundant rootdrip and horizontal lines of calcrete.

-- abrupt, smooth, unconformable contact with a stoneline --

Unit 3. "Riverton II Silt", 0.4 to 1.3 m, weakly cemented, pale brown (10 YR) sandy silt (nos. 310, 309) with isolated pebble lines and root drip. The upper surface is the eroded base of a calcrete horizon. Weak horizontal stratification is typical with dips changing from near 0° at the top to 4° away from the modern channel at the base.

-- abrupt, smooth, unconformable contact dipping at 3° - 4° away from the modern Vaal --

---

Vaal-Harts interfluve, and no similar source of dissolved carbonates and silica is available. The deposits at Wedburg are also perched on an andesite ridge, that still bears a veneer of shale.

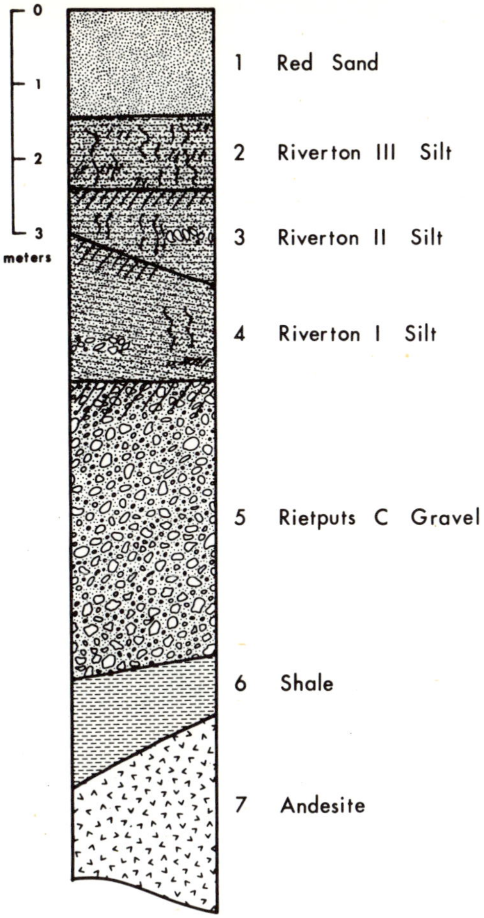

Fig. 5-14. Generalized sedimentary profile at Windsorton II.

Unit 4. "Riverton I Silt", 1.2 to 1.8 m, very pale brown (10 YR) sandy silt (nos. 308, 307). The upper 0.5 to 0.6 m is weakly calcreted with abundant root casts and calcrete lenses. Horizontal stratification dipping at 3° - 4° away from the modern river is well developed.

-- abrupt, smooth, unconformable contact --

Unit 5. "Rietputs C", 2.5 to 4.0 m, typical, calcreted massive gravels with white (10 YR) matrices of inhomogenous sand.

-- abrupt, smooth (undulating), unconformable contact --

Unit 6. "Bedrock", striated Ventersdorp andesite locally veneered by Dwyka shale.

The Windsorton II section records typical, calcreted, Rietputs-C gravels below elements of Riverton Members I, II, III, and IV. The exposure is particularly interesting because it records (at least indirectly) the separate calcretion of Riverton III, II, I and Rietputs C in one outcrop. Compared to Windsorton I, this exposure is dominated by channel-fringe and flood silts, without the high colluvial component at Windsorton I. The bedding dips in Riverton I and Riverton II suggest a repeated, backset levee situation, while the Riverton III stratification suggests sedimentation in a peripheral, ephemeral flood basin. The Riverton IV sands are dominantly colluvial, with local alluvial reworking.

The Windsorton I and Windsorton II provide a theme for categorization of the deposits along the right bank, south of the bridge. A cross-sectional illustration is presented in Fig. 5-15. However, expectable, additional facies may be noted. As on Riverview Estates, there are two facies of Riverton IV gravels -- the earlier colluvial variety with yellowish red (5 YR) matrices, and the latter Vaal variety with brown (10 YR) matrices. Both are best developed near the modern channelway. In addition there are both colluvial and alluvial gravels belonging to Riverton III, generally weakly calcreted with 5 Y, 10 YR white matrix colors.[11] Unfortunately the

---

[11]At two sites Riverton III gravels overlie more than 4 meters of Riverton I and II silts and sands. In the area east of the cemetery, Riverton III gravels reach the +9 m terrace surface

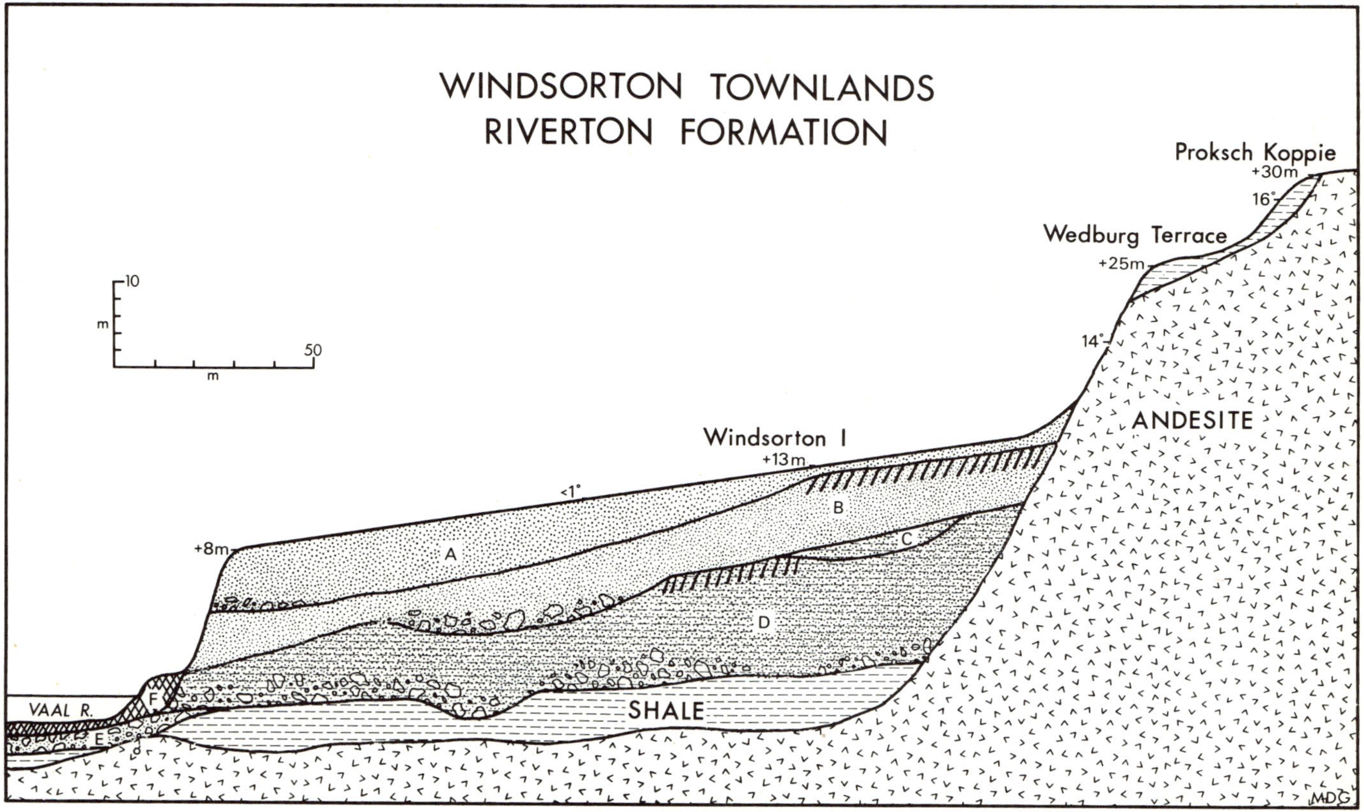

Fig. 5-15. Riverton and Rietputs formations at Windsorton; a cross-section from Windsorton I to the Vaal. A--Riverton IV red sandy colluvium; B--Riverton III sands; C--Riverton II silts; D--Riverton I silts; E--Rietputs C; F--Modern sediments.

terrains along the river have been intensively altered by diamond diggers, and primary deposits and morphologies are highly fragmented. The best exposures of Riverton IV alluvia are on The Island. Here up to 8 m of Riverton IV (Vaal) silts and sands cover Rietputs Member C.

North of the bridge, on the Windsorton side of the Vaal, the valley fill has been intensively mined and only minor fragments of the primary deposits persist. The available deposits are dominated by gravels of Rietputs B (weakly calcreted) covered by a veneer of reddish colluvial sands (Riverton IV). Elements of other Riverton Formation members were undoubtedly also present, but no record is available today. Andesite is the dominant bedrock.

The bedrock conformation beneath the valley fill at Windsorton and Riverview Estates was important in the disposition of the several aggradational and erosional phases. In general the bedrock surface undulates gently beneath the Rietputs Formation, but at some localities there are some spectacular features. Foremost among these is a gently rounded depression in the andesite bedrock which contains a 30 m thickness of gravels. The gravels in this depression, which is located along the east bank, north of the

---

and directly overlie Riverton I gravels. Van Riet Lowe (1937, 1952) categorized gravels following upon the Younger Gravels as "Youngest Gravels", identified in minor tributaries and erosional gulleys leading down to the Vaal, particularly along the Vaal's middle reach, near Vereeniging. Partridge and Brink (1967) also noted such deposits near Windsorton, labeling them "Gully Wash Gravels". Such gravels occur widely in the tillite and andesite terrains of the Lower Vaal Basin. In most cases such gravels are no more than shallow hillslope deposits, and in all cases they are locally derived. It is clear that some of these deposits correlate with Older Gravels and Rietputs Formation events, but most of them correlate with aggradation of the Riverton Formation, particularly Members III and IV.

highway bridge, extend to the top of the +12 m terrace, but they also extend 18 m below the modern river level. While the origin of this depression, which is called a "pot hole" by the diamond diggers, is problematical, it seems impossible that the erosion of such a feature in igneous bedrock could have been accomplished by late-Cenozoic fluvial processes.[12] Perhaps the best explanation is that this feature is the product of erosion along a fracture zone by a hydraulic vortex at the base of a late-Paleozoic glacier. Later in the glaciation the depression was filled by Dwyka sediments, which in turn were eroded out prior to the aggradation of the Younger Gravels. However, even the excavation of the softer Karoo sediments from the "pot hole" required extraordinary hydraulic forces (see Shepherd and Schumm, 1974, for an analogue). Another curious feature is a linear depression in the andesite, which runs eastward from the "pot hole" toward Riverview I. Since this depression is filled not only with gravels, but also with Dwyka sediments, it is best identified as a structural cleft in the Pre-Karoo Surface. Presumably this depression was followed by the Vaal during the initial aggradation of the Rietputs A. Other bedrock irregularities at Riverview Estates include several pedestals of shale which rise to near the top of the +12 m terrace (as noted above) as well as possibly shale-lined channels (see Söhnge and Visser, 1937).

By analyzing the shapes in the bedrock surface and the locations of high-energy sedimentation is it possible to speculate on main channel routes during aggradation of the valley fill at Windsorton (see Fig. 5-3 for reference). The gravels of Rietputs A have been repeatedly scarred by later intervals of floodplain incision. The remaining fragments suggest a progressive channel shortening from

---

[12]Potholes matching the standard definition of a spherical depression eroded by spinning cobbles (up to 30 cm in diameter), also occur in the some andesite channelways in the Lower Vaal Basin.

(1) a large loop extending from the "pothole" toward Riverview II and returning to the modern channelway near The Island, to (2) a relatively straight reach running from 50 to 150 m east of the modern channel. Given the bedrock control on channel location above and below Windsorton by the andesite ridges at Wedburg and Amandelhoogte, the large area of Rietputs-A sedimentation seems beyond the capacity of a progressively splaying Vaal channelway. Much of the gravels of Rietputs-A fill may be due to lateral erosion and subsequent reworking of a formerly much more extensive, Wedburg Terrace sedimentary fill. In addition, there may have been a major tributary stream draining a large region to the east of Windsorton.[13] Rietputs B sedimentation suggests a similar, although shorter, channel progression, looping initially across the "pothole" area and Riverton II, but eventually straightening out to run east of the modern channelway. As noted above, the modern channel was established during Rietputs-C sedimentation and all the Riverton Formation channelways, both aggradational and erosional, were located near the modern channel.

Chronology of the Rietputs and Riverton Formations at Windsorton

Combining the stratigraphy of the exposures described above with the additional facies represented around Windsorton, it is possible to construct an impressive history for the valley fill at this locality (schematic illustrations appear in Helgren, 1977; 1978).

---

[13]Support for the former existence of an important Vaal tributary joining the Vaal from the east near Windsorton is also provided by Rietputs-B, sandy facies sedimentation at Riverview II as well as by east-west gully-shaped unconformities in the valley fill on Riverview Estates. This tributary stream may have been important in deflecting the Vaal's course toward its modern axis. The absence of Rietputs A on the Windsorton side of the Vaal also is intriguing. It may only be a matter of no preservation, but it is also possible that a ridge of Dwyka sediments persisted near Windsorton until after aggradation of Rietputs A.

1) Abandonment of the Wedburg Terrace and incision of the Karoo sediments to irregularly expose a pre-Karoo valley. The floodplain-margin sediments were silts and sands of the "Basal Fines" at Riverview I. River levels: from +21 m to somewhat below present level.
2) Aggradation of Rietputs Formation, Member A, a complex valley fill of coalescing and merging mid-channel bars, combined with tributary gravels from coeval hillslope erosion. River levels: from below present level to above +12 m.
3) Floodplain incision, followed by development of the rubefied weathering zone on Rietputs A. Still later, the eroded gravel surfaces in the incised channels in the eastern sector of the valley fill were calcreted. River levels: from above +12 m to near present level.
4) Aggradation of Rietputs Formation, Member B (with gravel and sand facies), coincident with local erosion of Rietputs A and stripping of surficial sediments from the surrounding slopes, to be re-sorted in the channelways. Repetitive calcretion of sediments in and along the channels of the eastern valley-fill area. Calcretion also followed alluviation. River levels: from present level to at least +12 m.
5) Fill cutting and then aggradation of Rietputs Formation, Member C along and near the axis of the modern Vaal. River levels: from above +12 m to below the present level. Followed by calcretion.
6) Aggradation of gravels, sands, and silts (alluvial and colluvial facies) pertaining to Member I, Riverton Formation. River levels: near the present to about +8 m.
7) Fill cutting. River levels: about +8 m to about +4 m. Calcretion of surface of Member I.
8) Aggradation of silts (dominantly alluvial facies) of Member II, Riverton Formation. River levels: about +4 m to about +6 m.

9) Fill cutting. River levels: about +6 m to below present level. Calcretion of surface to Member II.

10) Aggradation of Member III, Riverton Formation (large-scale valley colluviation and alluviation, multiple sedimentary facies). River levels: below present to above +12 m.

11) Fill cutting. River levels: greater than +12 m to near present. Massive calcretion of Riverton III.

12) Aggradation of Member IV, Riverton Formation, initially yellowish red and reddish yellow sand and gravelly colluvia, followed by brown silty to sandy alluvia. River levels: near present to above +8 m.

13) Fill cutting. River levels: greater than +8 m to near present. Inceptisol develops on Riverton IV. (Riverton V not preserved along the Vaal at Windsorton.)

Younger Gravels at Barkly West

The Rietputs Formation is poorly preserved between Windsorton and Barkly West, and no equivalent terraces occur. The valley fills in this sector are dominated by fine-grained facies of the Riverton Formation, which are particularly well exposed in many tributary gullies. The typical inhomogeneous gravels of Rietputs C are subcontinuous near the Vaal channelway. Other, thin Younger Gravel beds occur sporadically beneath the Riverton Formation, but stratigraphic identification is not possible.[14] The most promising area for gravel investigation is The Bend, where the Vaal makes a long, southward loop (see Figs. 5-16, 5-17; Helgren, 1978). This looping channel plan appears to have developed progressively during

---

[14]Due to the lower channel gradient in this sector, fine-grained (non-rudaceous) facies of the Rietputs Formation could be expected. Such deposits would have been less durable in regard to succeeding erosional phases and would not be excavated by diamond diggers.

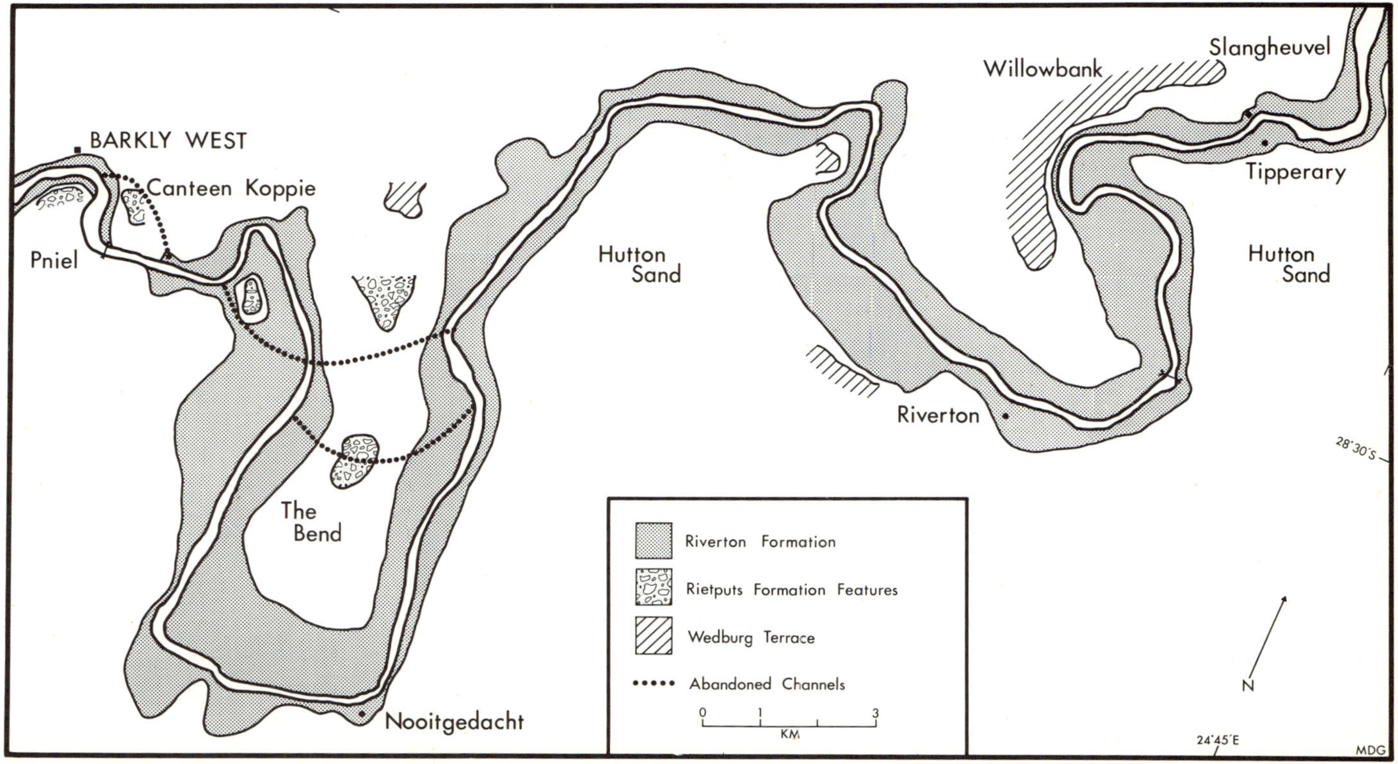

Fig. 5-16. Rietputs and Riverton formations near Barkly West.

Fig. 5-17. Vaal channel at the south end of The Bend. Glacially stream-lined andesite terrain in foreground. (Photo: K.W. Butzer)

Rietputs Formation times. The Rietputs-A channel ran across the neck of the loop, where an eroded, +14 to +16 m bench of Dwyka shale with a thin veneer of gravels is preserved. The Rietputs-B channel ran midway across The Bend, as evidenced by the presence of thick gravels at depth (see du Toit, 1907, map). The route of the modern channelway was established before aggradation of Rietputs C. In addition, thick gravels are present west of the river beneath the Riverton Formation fill near the Pniel mission ruins. Unfortunately these gravels were not exposed during this investigation. The modern terrain on The Bend is dominated by subdued, vegetated, eolian slope forms where the reddish yellow colluvia of Riverton III and IV have been worked into dunes on at least two occasions. East-northeast-trending, transverse ridges are visible on the aerial photographs.

The Younger Gravels terrace reappears at Barkly West, where the Vaal re-enters a deep, pre-Karoo channel. Relief exceeds 60 m in the adjacent, cobble and boulder-covered andesite terrains. The gravels terrace is best developed above the right bank of the modern Vaal at Canteen Koppie, but another residual is found further downstream, on the opposite bank at the site of the former diamond-rush camp of Pniel (Figs. 5-16, 5-18). Both areas have produced extraordinary quantities of artifacts and fossils (Helgren, 1978). Gravels also occur in the channel of the modern river.

The bedrock conformation near Barkly West is dominated by exhumed Paleozoic forms, which include micro- and macro-scale glacial valleys (trending 70°-85° E of N and 5°-15° E of N). These are superimposed on macro-scale, erosionally widened, lineament valleys, some of which are occupied by the modern Vaal channel (see Chap. 3 above). Dwyka sediments persist locally along the valley bottom and form a thin veneer over a broad area of the Pre-Karoo Surface northeast of Barkly West. Most important for interpreting Younger Gravels deposition is an abandoned channelway, along a widened lineament (Fig. 5-16). This channel appears to have been the Vaal's course during the first phases of Younger Gravel deposition,[15] with the modern channelway gaining dominance before aggradation of Rietputs C.

Barkly West was the first important center of alluvial diamond digging, and only fragments of the original deposits have survived. The foremost digging area was the +16 to +17 m terrace at Canteen Koppie (Figs. 5-16, 5-18). Here the gravels are veneered by 0.2 to 4 m of yellowish red to reddish yellow (5 YR) sand and silty sand, perhaps recording two intervals of valley colluviation (Riverton III

---

[15]This channel is still used by the modern Vaal when flood stage exceeds +8 m; it was also noted by Söhnge and Visser (1937).

Fig. 5-18. Bend in the Vaal at Barkly West. Canteen Koppie is at the left center (light tones) and Pniel is at the upper right. The Dikbosch Fracture cuts across the penninsula at upper left; there are no signs that the Vaal ever occupied this route.

and early Riverton IV). These colluvia include local pebbles and cobbles of andesite from the adjacent slopes. The gravels (up to 12 m thick) are typical pebble to boulder clasts in inhomogeneous, sandy matrices. However a more angular component is present, presumably reflecting fragments from erosion of nearby slopes. Fine-grained lenses occur at depth, as do discontinuous zones of calcretion. The surface of the gravels undulates, often intersecting the terrace surface. The upper part of the gravels preserves a rubefied cambic paleosol as well as thin layers of oxidized, fine-grained sediments. Based on this paleosol and the morphostratigraphic position as oldest fill of the modern valleyway, these gravels are correlated with Member A of the Rietputs Formation at Windsorton. These gravels were aggraded when the Vaal was using the abandoned channelway. Today Canteen Koppie is a national monument and no digging has been allowed for many years.

Across from Barkly West are the Pniel diggings,[16] where so few primary deposits are preserved that reconstruction of the stratigraphy is hazardous. At the surface are up to 4 m of red, sandy colluvium, as at Canteen Koppie. The gravels beneath have typical Younger Gravel textures and lithologies, with calcreted matrices (Söhnge and Visser, 1937). However, no terrace morphology is preserved and correlation to any member of the Rietputs Formation remains speculative. The gravels in the modern river at Barkly West were not exposed during this investigation, but correlation with Rietputs Member C must be inferred. In general, the Riverton Formation is poorly exposed and preserved near Barkly West. The dominant feature is a +8 to +9 m terrace of brown silts and sands pertaining to Riverton IV.

The Rietputs Formation from Gong-Gong to Winter's Rush

In the incised, pre-Karoo channelway below Barkly West, the modern Vaal descends nearly 50 m in the 18 km to Gong-Gong. In the adjacent terrains local relief exceeds 40 m and the steep slopes are veneered with rounded, andesite cobbles and red, sandy lithosols. Fills of Dwyka sediments persist in some gullies and along isolated channel fringes. Minor, fragmentary Younger Gravels deposits, either in the channel or along the valley footslopes are nearly continuous in this sector. However the deposits have been intensively mined, and correlation in the channelway is hazardous beyond

---

[16]Pniel diggings are located on Pniel Estates, originally founded by the Berlin Missionary Society in the 1840's. Pniel Estates has over 45 km of river frontage, extending from The Bend to beyond Waldeck's Plant. Consequently the toponymn "Pniel" is potentially ambiguous.

Rietputs Member C. The largest deposits of Younger Gravels were at the Good Hope and Bad Hope diggings (see Fig. 1-4), sites suggesting settings where valley alluvia merged with tributary fans. Elements of the Riverton Formation are also present in this sector, but the narrow width and steep gradient of the valley, as well as the steep confining slopes argue for limited primary deposition and for persistently poor preservation.

At Gong-Gong the Vaal's Paleozoic valley widens and a massive, valley-fill of Younger Gravels reappears (Figs. 5-19, 5-20, 5-21, 5-22, 5-23, 5-24). The deposits are preserved along the inside of valley bends at Gong-Gong, Waldeck's Plant, Longlands and Winter's Rush, and there is little doubt that the Younger Gravels were once continuous in this 13 km sector. Few deposits have survived the diamond digging, and during this investigation the remaining deposits were being rapidly consumed by large-scale mining. The most informative deposits are at Winter's Rush, and these provide a framework for ready categorization of the other surviving deposits.

The Winter's Rush deposit consists of four distinctive fills, which provide an extraordinary parallel to the features at Windsorton (see Fig. 5-20). The oldest fill consists of massive gravels with a rubefied paleosol on the surface; the second fill includes gravels with cyclically deposited sands and calcrete lenses; the third fill is dominated by silty sands and has a calcrete paleosol at the surface; and the fourth fill comprises brown, sandy silts. The correlation is immediate to Rietputs A, Rietputs B, Riverton Formation Members I and III, and Riverton Member IV respectively. In addition, Rietputs C is present at the base of the modern Vaal channelway. While the fills at Winter's Rush are locally superimposed, the characteristic facies are best developed in adjacent areas, and description by fill unit, rather than by type-exposure seems appropriate.

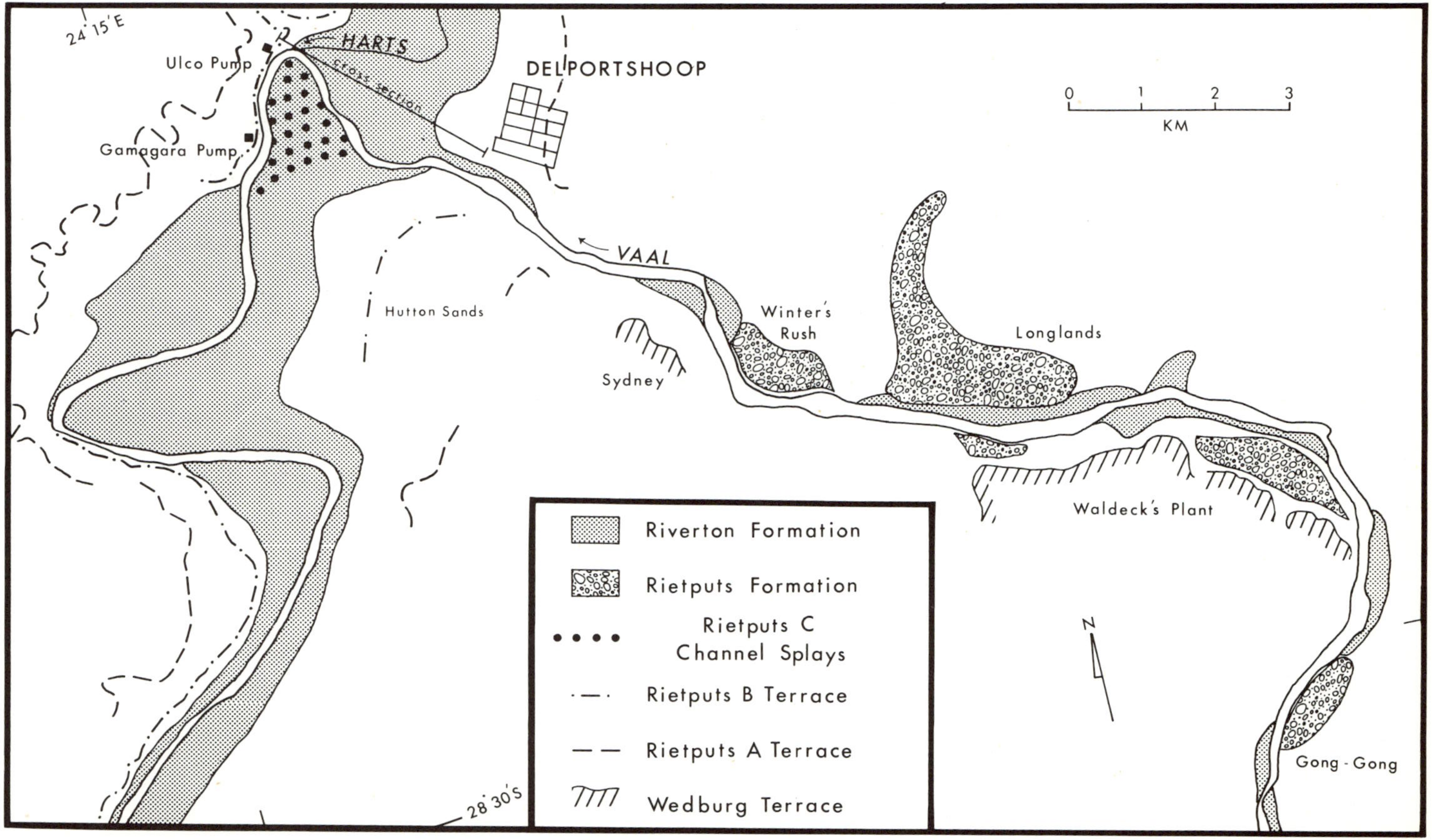

Fig. 5-19. Rietputs and Riverton formations, Gong-Gong to Delportshoop.

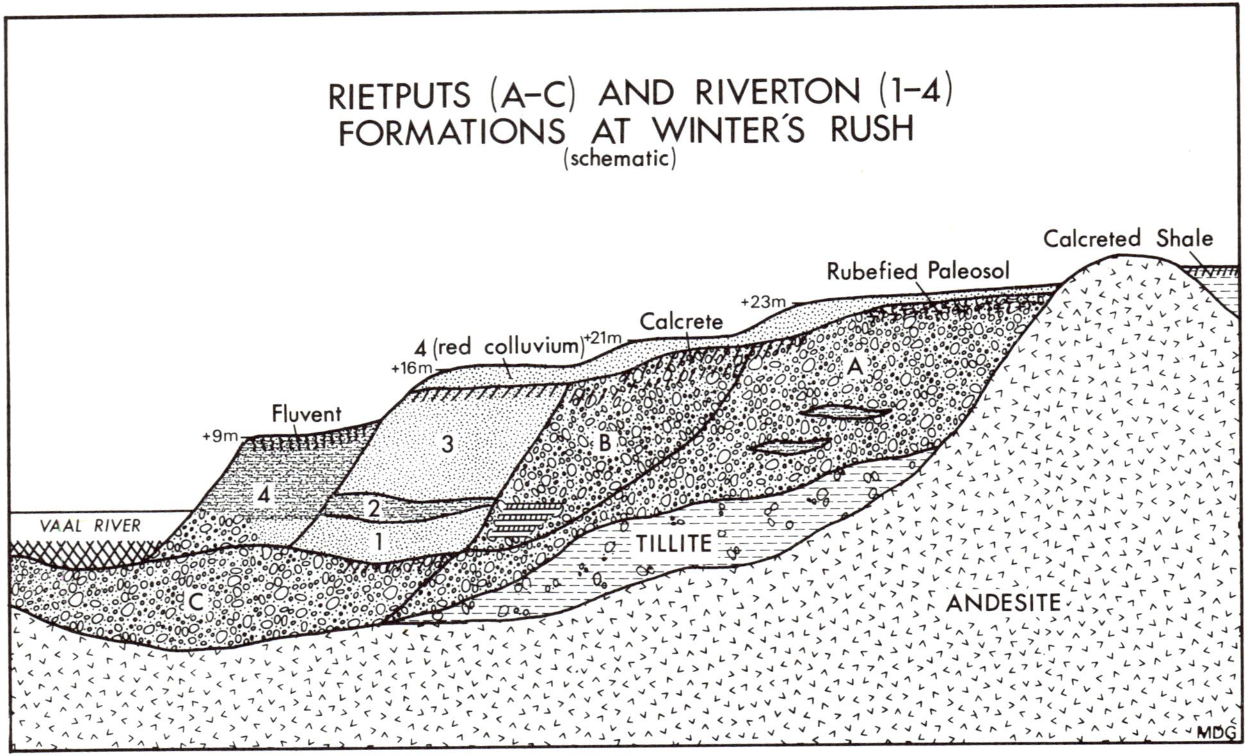

Fig. 5-20. Generalized cross-section of the alluvial fills at Winter's Rush.

Fig. 5-21. Exposure of Fill I and Fill II at Winter's Rush, near the center of the original valley fill.

Fig. 5-22. Lower Fill II at the downstream margin of the valley fill at Winter's Rush. Staff has 10 cm stripes.

Fig. 5-23. Vaal channel at Winter's Rush (view is upstream; photo: K.W. Butzer).

Fig. 5-24. Vaal channel at Gong-Gong (view is upstream; photo: K.W. Butzer).

Winter's Rush Fill I -- Rietputs Member A -- forms a +22 to 23 m terrace which merges with the calcreted, shale plains along the north side of the Vaal valley. Exposed thicknesses range up to 10 m above the tillites and striated andesites on this valley shoulder. Total thicknesses in excess of 25 m were probably once present. Typical textures in Fill I are the usual subrounded, pebble to cobble-grade gravels in quartzose, sandy matrices with crude, horizontal stratification (Figs. 5-20, 5-21). Andesite is the most common gravel lithology, while shale fragments (tillite matrix) and diverse, siliceous rocks (former tillite clasts) are secondary (see gravel analysis in Appendix B). Dispersed at the base of Fill I are andesite boulders (derived from the tillite) with volumes greater than $5m^3$! Matrix textures vary from inhomogeneous sands to better sorted, silty sands (nos. 057, 058, 059), while matrix colors in the pale brown range (10 YR) are characteristic.[17] Also present are rare, dispersed lenses of finer sediments up to 30 cm thick. These include fluvial, yellowish red and red (2.5-5 YR), silty sands (no. 060), and floodplain soils of pale brown (10 YR), vertic, clayey silt and pink (7.5 YR), blocky, silty sand (nos. 100, 101). At the surface of Fill I is a rubefied paleosol with red and yellowish red (2.5-7.5 YR) matrix colors. In areas where the surface of Fill I was scoured before deposition of Fill II, the upper gravel matrix is often indurated, with pinkish white (5 YR) calcrete (no. 061). At the margins of the valley, Fill I is often veneered with a meter of pale pink (10 R) calcrete (secondarily cemented silty sand, belonging to Member III of the Riverton Formation, no. 070), and up to 2 m of massive, red, colluvial and eolian sand (referable to early Member IV of the Riverton Formation, no. 071). Correlated with Fill I are the calcreted, coarse gravels located in a largely unexposed, tributary channelway that is several tens of meters wide, nearly 10 m deep, and extends at least 5 km to the north of Winter's Rush and Longlands.

Winter's Rush Fill II -- Rietputs Member B -- occupies scoured channelways in the surface of Fill I and forms a separate +19 to 20 m terrace. Thicknesses typically vary from 3 to 10 m, and both channel and overbank deposits are preserved. In general, the channel sediments of Fill II "coarse upward" with a lower zone of relatively fine sediments and an upper zone of coarse gravels. An examplary mid-channel facies of the lower zone located near the center of the Winter's Rush deposits totals 1.5 m and consists of complexly-bedded lenses (up to 40 cm) of sand, silty sand, and clayey sand (Fig. 5-21). Dispersed clasts up to 30 cm across occur

---

[17]Matrix colors at the base of the deposit locally are pale yellow (5 Y), from oxidation related to recent watertable fluctuations.

throughout the facies, as do conglomerates of subrounded, medium and fine pebbles. Both planar and trough cross-stratification are present, in addition to the dominant, crude horizontal stratification. The sand fractions include clasts of previously-cemented gravels matrix, and many strongly-weathered, quartz fragments and splinters. These lenses are also variably cemented, ranging from loose and uncemented, to extremely-cemented/indurated (nos. 062, 063A, 063B, 064, 066a, 066b, 067). The predominant colors in this example of Fill II are in the very pale brown (10 YR) range, but important areas are also white (10 YR), light red (2.5 YR), and yellowish brown (10 YR).

A channel-margin facies of the lower zone of Fill II totals 2.8 m in an exposure near the downstream edge of the Winter's Rush deposits and contains a comparable diversity in sediment types (see Fig. 5-22). At this site 1.6 m of horizontally stratified, pale brown (10 YR), coarse sands with dispersed pebbles and cobbles (no. 330) is found above the weakly calcreted gravel surface of Fill I. These sands are capped by 1.2 m of alternating layers of light gray (3.5 Y), prismatic, clayey sand silt (sample 331) and pink (7.5 YR), structureless sand (no. 332). There are six layers (5 to 25 cm thick) of each sediment type, and at the surface there is a 15 cm layer of white (8 N) calcrete (indurated silty sand; no. 333). Also present near this site are over 3 m of overbank sediments of Fill II age. This sediment is a vertically structured, pink (7.5 YR) silty sand (no. 280) with calcrete "poupées" (hollow concretions) dispersed throughout. Comprising the upper layer of Fill II are the typical coarse gravels, again with pale brown (non-calcreted; 10 YR) or white (calcreted; 10 YR) matrices (nos. 068, 069). The transition between the upper and lower zones of Fill II usually is abrupt. The upper Fill II gravels, up to 8 m thick, are monotonous throughout the Winter's Rush area and are locally capped by the same calcrete and red sands that cover Fill I.

Younger Gravel C -- Rietputs Member C is present near and in the modern Vaal channel at Winter's Rush. Fragmentary exposures and interviews with diggers indicate the typical, coarse gravel textures with thicknesses up to 6 meters. The upper surface of these gravels is calcreted along the channel periphery.

Winter's Rush Fill III forms a +15 to 16 m terrace. The related sediments are nearly identical to those of the Windsorton I and II exposures described above. At the base are pale brown (weakly calcreted), fluvially bedded sands belonging to Member I of the Riverton Formation. At the top of the terrace is a thick horizon of pale brown to reddish yellow, colluvial sands with a Member III calcrete paleosol at the surface. In addition, red colluvial sands of early Riverton IV locally cap the calcrete.

Winter Rush Fill IV forms a +8 to 9 m terrace along the modern Vaal, and is dominated by the brown, sandy silts typical of Member IV of the Riverton Formation. At the base of the fill are gravels with brown matrices, apparently reflecting Younger Gravels reworking during the early stages of deposition. A low, brown silt terrace referable to Member V of the Riverton Formation may have been present at Winter's Rush before the diamond rush.

The chronology of the Younger Gravels at Winter's Rush is essentially the same as at Windsorton:

1) Erosion of the Dwyka sediments then filling the present river valley and abandonment of the Wedburg Terrace. River levels: from above +30 m to below present.
2) Aggradation of the Fill I (Rietputs A) as a complex of coalescing lateral and longitudinal braid-bars. River levels: below present to about +23 m.
3) Incision of Fill I. Rubefaction of the Fill I surface. River levels: about +23 m to about +3 m.

4) Aggradation of Fill II (Rietputs B) complex within channels scoured into the Fill I surface. Local calcretion of the Fill I surface. River levels: about +3 m to about +20 m.

5) Incision of Fills II and I, with river levels dropping to near modern positions; modern channel position established and the gravels of Rietputs C deposited.

6) Later still the multiple cut-and-fill episodes of the Riverton Formation also occurred in the Winter's Rush area.

The fragmentary Rietputs Formation deposits at Gong-Gong, Waldeck's Plant, and Longlands are compatible with the apparent sequence of events at Winter's Rush. The modern Vaal descends 12 m between Gong-Gong and Winter's Rush. At Gong-Gong a single +23 m terrace contains both Rietputs Members A and B (Fig. 5-24). Rietputs A is the typical, massive gravel body which locally exceeds 30 m in thickness and bears a rubefied paleosol. Rietputs B is preserved in a narrow, 30-meter wide channelway, cut along the right edge of Rietputs-A fill. The sediments are calcreted coarse sands with multiple gravel lenses, and massive calcreted gravels; thicknesses exceed 20 m. Rietputs Member C is continuous in the river channel. At Waldeck's Plant the remaining Younger Gravels deposits are exclusively referable to Rietputs A which forms a +22 m terrace with a rubefied paleosol at the surface. At Longlands even fewer than usual primary deposits have survived the century of diamond digging. Remnants of a +22 to 23 m Rietputs Formation terrace are present, and deposits along the right margin are generally calcreted, but rubefied paleosols are also common.[18] A +9 m

---

[18] Several exposures at Riverview Estates and Longlands suggest that Member A of the Rietputs Formation may have been calcreted before the characteristic, rubefied paleosol developed on its surface. Accurate discrimination seems unlikely in any case. The early members of the Riverton Formation are particularly rare in the Gong-Gong to Winter's Rush sector. The erosional cycle following

terrace at Longlands records both colluvial (red sand with gravels) facies of Member IV of the Riverton Formation. The Younger Gravel record between Gong-Gong and Winter's Rush will soon disappear due to continued mining.

The Rietputs Formation At The Vaal-Harts Confluence

At the Harts confluence the Vaal enters the Paleozoic "Kaap Valley" (du Toit, 1910; see Chap. 3 above) and strikes southwestwards toward the Orange. Here the Vaal is at an elevation of 1000 m, having dropped over 100 m from its elevation at Windsorton. In the Kaap Valley the bedrock constraints placed on the development of the Rietputs Formation also change, even though the channel remains broadly confined to a pre-Karoo channelway. At Windsorton and downstream of Barkly West the periods of incision preceding and interrupting Rietputs Formation sedimentation repeatedly stopped near the base of the pre-Karoo, andesite valleys. But downstream of Winter's Rush the primary period of incision frequently stopped before the base of the "soft", Karoo sediments was reached. As a result, the later periods of floodplain incision during the Rietputs Formation interval could and frequently did erode deeper than the earlier incision, so that descending terrace sequences provide a morphologic theme to beyond the Orange confluence.[19] Near the Vaal-Harts confluence, eroded shoulders of the Rietputs-A terrace are found at +24 to 27 m, Rietputs-B terrace at +18 to 19 m, while Rietputs C is found in a large area near the modern Vaal-Harts confluence.

Riverton III appears to have been particularly thorough. A minor Younger Gravels deposit occurs opposite Longlands at Niekerk's Rush. Correlation of this deposit is unclear.

[19]Below Winter's Rush the number and extent of rudaceous Rietputs Formation outcrops diminishes severely. Consequently sediments associated with the terraces are rare, and if present, are poorly exposed.

The area of the Vaal-Harts confluence is extraordinary also because it is a zone where three different sets of sediments interdigitate. First are the coarse-grained sediments of the Vaal River; second are the finer alluvia of the Harts River; and the third are limestone tufas which intermittently aggraded along the front of the Gaap escarpment and occasionally reached the Vaal and Harts rivers. The interdigitation of these three sediment suites during Rietputs Formation sedimentation has produced a complex but predictable array of facies variations.

In addition, the location of the Rietputs Formation deposits near Delportshoop is strongly controlled by local undulations in the Pre-Karoo surface. Major NE-SW-trending ridges are found on both sides of the river. The highest extends from Borrelkop (north of Delportshoop) across the Sydney village. The other is more subdued, starting west of the Ulco pump and slowly rising southward. Within this valley the Rietputs and Riverton formations have been compressed against each other, and each intervening period of incision destroyed large parts of the preceding terrace and associated deposits. However, parts of all three are preserved on both sides of the rivers.

Rietputs Formation deposits and planation features are preserved on the Delportshoop townlands, on the farm Sydney, and along the right bank of the Vaal-Harts channel (Fig. 5-19). The largest area of deposits are on the farm Sydney, but unfortunately these are essentially unexposed. Nearly a century of diamond digging at Sydney was terminated in the 1960's and most of the digging areas were reclaimed for capital-intensive agriculture. The Sydney deposits are irregularly buried beneath a veneer of red Hutton Sands (see Chap. 2 above), but the smoothing contours still indicate deposits of the Rietputs Formation at depth. The walls of the few pits still available have slumped badly, and only fragmentary outcrops are visible. Consequently this report emphasizes the deposits across the Vaal from Sydney.

No rudaceous facies of Rietputs Member A are now exposed near the Vaal-Harts confluence. The Rietputs-A record consists of terraces and a massive limestone tufa fill. The former are sporadically exposed as an eroded valley shoulder (at +24 to 27 m) all along the west margin of Sydney (Fig. 5-19), and best demarcated by a low, calcreted scarp on pockets of Dwyka sediments. This terrace occurs at the same level on the Delportshoop townlands, where it is preserved as a calcreted shale plain (the basal part of a pediment-form footslope), grading eastward into the Rietputs-A shale plain north of Winter's Rush. On the west side of the Vaal-Harts, the Rietputs-A terrace is little more than a gentle slope inflection on the limestone tufa plains. An 18 m thick limestone tufa of Rietputs-A age is exposed along the Rietputs-B terrace scarp between the Ulco and Gamagara pumps (Figs. 5-25, 5-26, 5-27, 5-28). The exposure of these materials at the Ulco pump is exemplary.

ULCO PUMP

Unit 1. (top) "Tufa", while deposited during the Rietputs-A sedimentation, the weathered and corroded top of this unit was planed at +18 m during the deposition Rietputs B and demarcates the Rietputs-B terrace. (a) The upper 3.5 m of the remaining deposit consist of white (N), variably cemented (usually indurated), cryptocrystalline limestone (limestone tufa), which has a texture after HCl treatment of clayey silt sand (no. 128). (b) The middle 0.5 m consist of white (7.5 YR), indurated, argillaceous sandstone, which has a texture after HCl treatment of fine sand (no. 130). (c) The lower 12.0 m consist of white (N), variably cemented (usually indurated), cryptocrystalline limestone, which has a texture after HCl treatment of clayey sand silt (no. 129). All these subunits have crude, horizontal bedding, while rare, dispersed, rounded pebbles and cobbles occur in subdivisions (a) and (c). Micro-solution cavities

Fig. 5-25. Vaal-Harts confluence (photo: K.W. Butzer)

Fig. 5-26. Vaal-Harts confluence from the Ulco pump. The Harts (left) is aggrading a spit into the Vaal channel.

Fig. 5-27. Ulco pump terrace at the Vaal-Harts confluence.

in this unit, which were re-filled with tufa during the primary aggradation, indicate that some karstic solution was pene-contemporary.

-- abrupt, wavy unconformity --

Unit 2. "Bedrock", black to very dark blue Dwyka shale, which forms a bench +2 m above the present river level.

The Ulco pump deposit is a floodplain fill largely composed of finely bedded limestone tufas. These tufas were precipitated from $CaCO_3$ - saturated solutions draining the 11 kilometer-distant Gaap Escarpment (Butzer et al., 1978).[20] The detrital fractions come

[20]Depositional association with the tufa fan complex Ul-3 at Ulco must be assumed (Butzer, et al., 1978).

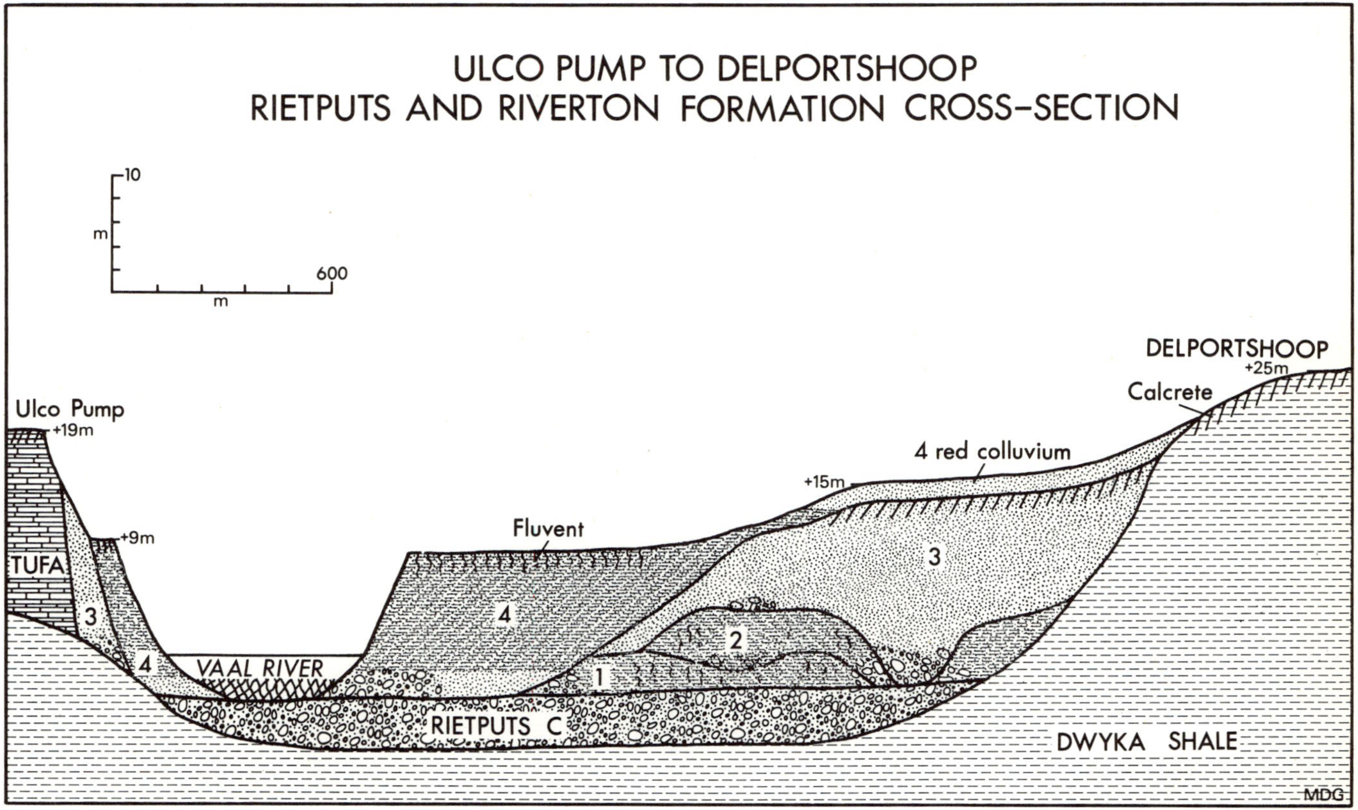

Fig. 5-28. Terraces of the Rietputs and Riverton formations at the Vaal-Harts confluence.

from the escarpment, as well as from the Harts and Vaal Rivers. The medial band of cemented sand in the described exposure probably is a mixture of tufa and Harts sands. Proceeding southward along the pump scarp, lenses (up to 1 m thick) of Vaal River gravels and Harts River silts become progressively more frequent in the tufa complex, and a formerly broad area of facies interdigitations was probably once present along the modern channel axis. It is unclear whether a systematic sequence of paleosols can be associated with the Rietputs-A terrace. The calcretes on the Sydney terrace residuals probably are nearly as old as the terrace formation, but the calcretes on the Delportshoop townlands may be of Riverton Formation age. The surface of the tufa west of the river typically is a corroded calcrete with secondary and tertiary recrystallization. However, in a pit near the Ulco shooting range (5 km from the river) an oxidized (limonitic) zone of tufa (over 2 m thick) is preserved beneath a relatively recent calcrete. It is tempting to correlate this oxidation zone with the rubefaction of Rietputs A at Winter's Rush and Windsorton.

The Rietputs Member A floodplain was later incised, and the tributary gullies deepened. Eventually renewed aggradation at the Vaal-Harts confluence culminated in the Rietputs Member B fill and the related +18 to 19 m terrace (Fig. 5-27). At Delportshoop, Rietputs B is represented by a minor, discontinuous, erosional bench notched into the Rietputs-A terrace. However at Sydney, Rietputs B forms a broad fill terrace with over 10 m of sediment. The fragmentary exposures on Sydney indicate up to 6 m of calcreted, fluvially bedded sands above the typical, calcreted, granule to coarse-cobble conglomerate. Unfortunately there are too few exposures to establish systematic relationships. Toward the end of Rietputs-B sedimentation, the Ulco Pump terrace was planed off at +18 to 19 m and the nearby erosional gullies filled with fluvial sediments. A residual is found at the Gamagara pump, where 8 m of indurated sands and gravels exhibit complex, fluvial bedding, including both planar and trough cross-stratification. These sediments bank unconformably against the earlier tufas and record the Vaal filling of a tributary channel. Aggradation of Rietputs B was not accompanied by tufa aggradation. Either these tufas did not have sufficient volume to reach the river valley from the escarpment, or they were no longer being deposited.

Following deposition of Rietputs Member B yet another interval of floodplain incision occurred, and most of the modern Vaal channel positions were established. The channel sediments of this period are Rietputs Member C, preserved beneath a thick cover of Riverton sediments. Rietputs C, which is up to 6 m thick and composed of the usual, irregularly calcreted conglomerate with inhomogeneous sandy matrices (no. 103; see gravel analysis in Appendix B), is spread across a large area (Fig. 5-19) and it is apparent that the Vaal channel lengthened significantly during this interval of limited deposition.

Elements of both Vaal and Harts river facies of the Riverton Formation are well preserved by three terraces near Delportshoop: a +14 to 16 meter terrace pertaining to Member III, a +9 to 10 meter terrace pertaining to Member IV, and a +4 to 5 meter shoulder along the channels referable to Member V (Fig. 5-28). Eroded remnants of Member I and II are present at depth.

Vaalian Riverton I sediments vary from 2 m of strong brown (7.5 YR), prismatic sandy silt (no. 297) to 0.8 m of yellow (10 YR), massive silty sand (no. 325). A Harts River facies totals 0.7 m and includes reddish yellow (5 YR), laminated sand (no. 105A); brown (7.5 YR) silty sand (no. 105B); and light brown (7.5 YR), cross-bedded clayey sand silt (no. 104). All facies have "root drip" and suggest multiple episodes of calcification.

Riverton II (Vaal) sediments vary from nearly 2 m of light brown (7.5 YR), blocky, clayey silt sand (no. 296) to 0.5 m of light brown (7.5 YR), faintly prismatic, clayey sand silt (no. 328). A Harts River facies totals 0.7 m and is composed of yellowish brown (10 YR), laminated, clayey silt (no. 106), which includes derived and abraded gypsum crystals.[21]

Member III of the Riverton Formation is dominated by massive sandy colluvia with a calcrete paleosol at the surface. Colluvial facies along both the Vaal and Harts are massive brownish yellow (10 YR), light brown (10 YR) and reddish yellow (10 YR - 7.5 YR) sand, silty sand, and sandy silt (nos. 329, 117, 116, 112, 111, 108). Vaalian, alluvial facies of Member III are poorly preserved, but fragmentary outcrops suggest inhomogeneous sands with gravel lenses. Harts River alluvial facies include light brown (7.5 YR), clayey sand silt (nos. 109, 110) and multiple, 10 cm layers of very pale brown (10 YR) marl. All of the Harts facies include derived gypsum crystals.

Across the top of the calcrete on Member III, on the +14 m terrace, and then continuing downslope to near the modern river level, is the reddish colluvial facies of Riverton Member IV. The dominant textures are sand and silty sand (nos. 114, 118); dispersed pebbles are common and there are up to 1 m thick horizons of reworked gravels, with red sandy matrices. Above these colluvia are brown to pale brown, sandy silt, silty sand, and massive sand of the alluvial facies of Member IV. These sediments comprise the +9 m terraces on all sides of the Vaal-Harts confluence as well as the fills in the tributary gullies within the Ulco pump terrace. At the base are reworked gravel horizons, up to 3 m thick, with brown silty matrices. During this Riverton IV phase, the Vaal

---

[21]The gypsum crystals in the Harts River facies of Riverton Formation Members II and III were apparently derived from the large gypsum deposit at Melkvlei, east of Delportshoop.

channel lengthened to assume its modern positions. Intersecting with earliest sediments of this alluvial phase of Riverton IV are minor limestone tufas from the Gaap Escarpment which fill the upper parts of tributary gullies in the Ulco pump terrace.

Riverton Formation Member V represents another alluvial episode, with minor, +4 to 5 m shoulders of brown sandy silts and silty sands embanked against similar sediments of the Member IV terrace (Figs. 5-29, 5-30).

The long term geomorphic trends at the Vaal-Harts confluence include continuation of the Vaal channel lengthening, begun during Older Gravels times, as the Vaal and Harts episodically continued re-excavation of the pre-Karoo valleyway. Patterns of erosion and sedimentation indicate close parallels to the chronology at Windsorton. The record of Rietputs Formation events at the Vaal-Harts confluence can now be summarized.

Chronology of the Rietputs Formation at the Vaal-Harts Confluence

1) Erosion of most of the Dwyka sediments filling the pre-Karoo valley near the modern confluence. River levels: above +35 m to +2 m.
2) Aggradation of Rietputs Member A to establish a floodplain near the +26 m level. This aggradation included rudaceous sediments from the Vaal River, arenaceous sediments from the Harts River, and argillaceous, carbonate sediments from springs along the Gaap Escarpment. River levels: +2 m to above +26 m.
3) Downcutting into the +26 m floodplain. River levels: above +26 m to near the modern river level.
4) Aggradation of Rietputs Member B to establish a floodplain near +18 m, resulting in planing off of the Ulco pump terrace at +18 m. The Vaal sediments were sands and gravels. No tufa aggradation. Harts River sediments are either not exposed or preserved. River levels: from near modern level to above +18 m.

Fig. 5-29. Harts River at the Vaal confluence, January 1973.

Fig. 5-30. Harts River at the Vaal confluence, November 1977. Riverton V terraces were scoured out by major floods in 1974 and 1975.

5) Downcutting into the +18 m floodplain, later followed by aggradation of Rietputs Member C. The Member C channel lengthened significantly in the confluence area, and generally established the modern channelway. River levels: above +18 m to below the modern level.

6) The five-fold cycle of floodplain aggradation and channel filling represented by the Riverton Formation. River levels: high positions recorded at +14 to 16 m, +9 to 10 m, and +4 to 5 m, as well as four low positions at or below modern river level.

The Rietputs Formation from the Harts Confluence to the Riet Confluence

In the 110 river km between the Harts River and the Riet River confluences, the modern Vaal River drops only 12 m. Along this route the two Rietputs Formation terraces occur subcontinuously on both sides of the river (Figs. 5-31, 5-32, 7-3), and gravels of Rietputs C can be verified within the modern channel as far as Schmidtsdrif. However the Rietputs Formation terraces frequently form erosional benches cut into Dwyka shales and tillites, rather than floodplain fills. Diamond digging is rare and sporadic. Consequently there is little sedimentological record of river behavior. In addition the east bank terraces are often buried under a thick mantle of Hutton Sands.[22] Nowhere in this stretch are there massive gravels comparable to those at Windsorton or Winter's Rush.

---

[22]With the addition of the discharge from the Harts River below Delportshoop, preservation of the Riverton Formation changes, and the erosional episode between Members III and IV appears to have been exceptionally vigorous, so that Riverton III terraces are comparatively rare. "Floodplain" surfaces are dominated by the colluvial and alluvial facies of Member IV, while Member III sediments dominate at depth.

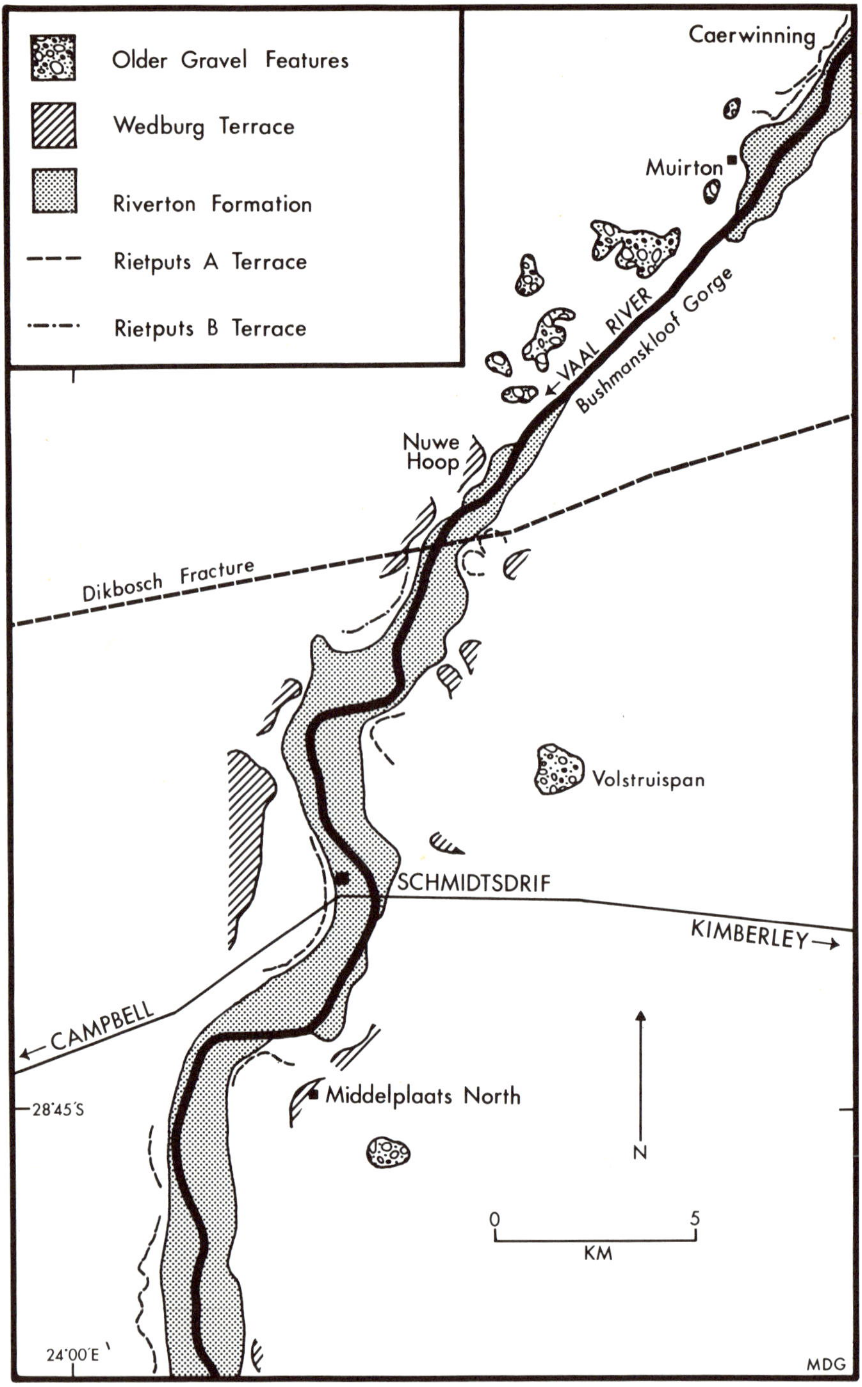

Fig. 5-31. Older Gravels, Wedburg Terrace, Rietputs terrace, and Riverton Formation near Schmidtsdrif.

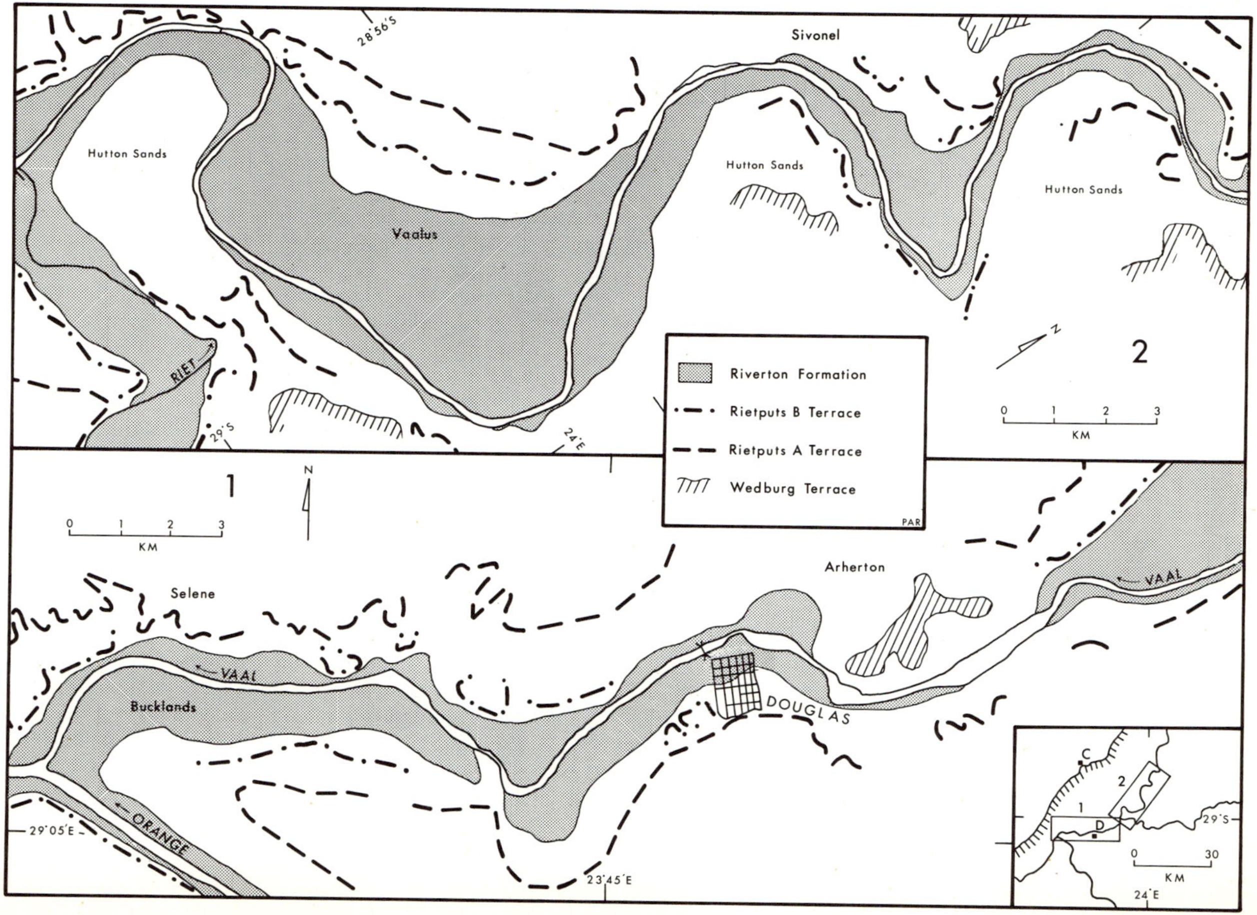

Fig. 5-32. Terraces along the Vaal River from Sinovel to Bucklands.

In this reach the Vaal bed lies near the floor of a pre-Karoo channel, still partly filled with Dwyka sediments, and generally centered along the erosional unconformity between rocks of the Ventersdorp and Transvaal Systems (see Chap. 3). The only major exception is the Bushmanskloof gorge, where the Vaal follows a pre-Karoo trough through a massif of Ventersdrop volcanics. The pre-Dwyka terrain continues to provide a relatively narrow river valley, bounded on the east by a descending, convex ridge of Ventersdrop volcanics, and on the west by a nearly continuous ridge of Transvaal sedimentaries. Within this valley the Vaal aggraded and incised its Rietputs Formation floodplains, leaving behind scars of these activities on the late-Paleozoic filling of Dwyka sediments.

Proceeding downstream from Delportshoop, the Rietputs Formation terraces are nearly continuous along the right bank to the opening of the Bushmanskloof gorge, while on the opposite side of the river, possible outcrops terraces are generally obscured by a thick veneer of Hutton Sands. On the farms Erindale and Caerwinning the terrace pattern mirrors the Harts confluence theme with a Rietputs-A terrace (at +26 m) of tufa fill and a Rietputs-B erosional bench (at +18 m). Two major Riverton Formation terraces (at +14 to 16 m and at +9 to 10 m) are also present.

Further downstream the only deep gravel exposure in this reach is preserved at Muirton.[23] In this fragmentary deposit a 1 to 4 m thickness of uncemented, granule to boulder-grade, basal gravels is succeeded upward by a 10 to 14 m thickness of variably cemented, coarse and fine sands with accessory rudaceous and argillaceous lenses that are dominated by high-energy, chaotic bedding. Correlation to Rietputs A and B respectively, is tempting. The sequence is capped with a meter of calcreted slopewash, with derived, Early

---

[23]Gravel beds are present, but unexposed, beneath Riverton sediments in several areas between Delportshoop and Muirton.

Stone Age artifacts, and a final 75 cm of light brown, uncemented slopewash, composed of clayey sand silts referable to Riverton Member IV (see also Helgren, 1978). This erosionally truncated gravel deposit is associated with a +14.5 meter terrace of Riverton age.

From Muirton past the Riet confluence, the ridges of Transvaal sedimentaries generally blocked the entry of limestone tufas into the Vaal valley during Rietputs A times, and the high terrace is usually only perserved as a bench cut into the Dwyka sediments. Below Muirton the Vaal enters the Bushmanskloof gorge. Rietputs Formation terraces do not reappear until the gorge widens at Nuwe Hoop. Here on the westbank the higher Riverton Formation terrace (at +16 m) is backed by poorly preserved terraces at +20 m and +25 m. Both these surfaces carry shallow, calcreted, colluvially-derived gravels. At Schmidtsdrif and across the river on Rooipoort (Volstruispan) the high terrace (at +27 to 28 m) is preserved as calcrete-coated benches on Dwyka sediments. South of Schmidtsdrif the loops in the river at Middelplaats North, Platdrif, Sinovel, Olie Rivier, Vaalus, and Malabar[24] all enclose the suite of Younger Gravels terraces (Figs. 5-31, 5-32). The highest terrace (at +22 to 24 m) most frequently is little more than a bench of calcreted shale or tillite, while the middle terrace (+17 to 19 m) typically is a non-cemented, erosional bench cut into Dwyka sediments. Both these benches often carry an irregular coating of pebble to cobble-grade gravels. Two lower terraces at +14 to 16 m and +9 to 10 m belong to the Riverton Formation.

---

[24]On Malabar, Sinovel, and Platdrif abandoned diamond diggers' pits are found on both the Rietputs A and B terraces. On Olie River and Middelplaats South, gravel beds of uncertain correlation are present beneath Riverton fill.

The Rietputs Formation near Douglas

Below the Riet River confluence the modern Vaal begins its descent toward the pre-Karoo gorge of the Orange River at Read's Drift. In the 26 river km between the Riet and Orange confluences, the Vaal drops 18 m and flows on or near rocks of the pre-Karoo surface throughout this distance. Downstream from Malabar, the Vaal follows the pre-Karoo channel through a breach in the Spioenkop Anticline at Arherton, and below Douglas it passes through another breach in the andesite bedrock at Blaauwboschdrift (Fig. 5-32). However, elsewhere in this sector Dwyka sediments rise above both banks of the river.

The Rietputs Formation terraces in this reach essentially replicate those at the Harts confluence. The highest terrace is a floodplain fill of massive tufas derived from springs along the Gaap escarpment (see Butzer et al., 1978). These tufas interdigitate with Vaal River rudaceous sediments. The Rietputs-B terrace, which is also associated with rudaceous Vaal River sediments, is demarcated by an erosional bench cut in the Dwyka sediments on either side of the river. Terraces of Riverton age are also present.[25]

The highest terrace of Rietputs Member A, is preserved on both sides of the river (Fig. 5-32). South of Douglas it is a poorly-defined, +33 m gravel and calcrete-coated bench on shale and andesite bedrocks. The surficial, calcreted gravels on this feature indicate significant bedrock stripping during the establishment of this terrace. However on the right bank (north) this terrace is preserved as an impressive scarp of limestone tufas. In some areas this scarp has retreated more than 2 km since the original tufa deposition, and consequently now stands at an unusually high elevation above the modern river (up to +60 m). However, on Selene,

---

[25]For a discussion of the terraces along the lowest Riet River, see Chapter Seven below.

where this scarp has suffered only limited retreat, it stands at +36 to 37 m (Fig. 5-33). Exposed on Selene are 9 m of sediments which record part of the depositional history of both the tufas and the coeval Vaal River.

SELENE

Unit 1. "Tufa", 7.0 m, crudely stratified, strongly-cemented, silty sands that include rare, dispersed pebbles. The colors vary from light gray (10 YR) and very pale brown (10 YR) to white (10 YR and 7.5 YR) and pinkish white (7.5 YR) (nos. 242, 243, 244). The fine sand fractions include weathered quartz and andesite fragments. The upper surface of the tufa has been extensively weathered and corroded to form solution cavities up to 0.5 m deep.

-- clear, smooth, conformable contact --

Unit 2. "Gravels and Sands", 1.1 m total. The upper 60 cm is horizontally stratified, white to very pale brown (10 YR), weakly-cemented, silty sand (no. 241) with frequent granules and fine pebbles. The lower 50 cm is planar cross-stratified, and consists of two, separate sediment wedges -- (a) a weakly-cemented, granule to small cobble conglomerate with a white to very pale brown (10 YR) matrix of fine and medium sand (no. 240); and (b) weakly-cemented, light brown (7.5 YR) sandy clay silt (no. 239). The lower section suggests a gravel dune moving over a fine-grained substratum.

-- clear, smooth conformable contact --

Unit 3. "Basal Gravels", 90 cm, weakly cemented, granule to cross-cobble conglomerate with crude horizontal bedding (see gravel analysis in Appendix B). The sub-rounded gravels contain a significant number of locally-derived, shale clasts. The matrix is very pale brown (10 YR) silty sand (sample 238).

-- abrupt, smooth, unconformable contact --

Fig. 5-33. Rietputs-A terrace at Selene. The Selene sediments form the caprock.

Unit 4. "Bedrock", more than 25 m of dark blue, Dwyka shale. The upper surface of the shale is planed at +27 to 28 m above the modern Vaal.

The Selene exposure provides a conformable, "fining upward" sequence, demonstrating that the Vaal was carrying medium and coarse gravels while the tufa terrace was aggrading. High proportions of amorphous silica cements are present in these sediments, indicating that the silicates of the local shales and igneous bedrocks were being weathered to introduce considerable silica into the groundwater. Although this exposure extends for over 200 m with only minor facies changes, there is no evidence whether the "fining upward" indicates decreasing river competence, or only that the channel axis migrated away from the site. The highly-corroded, upper surface of the tufa indicates that important intervals of weathering postdated the tufa deposition.

The lower Rietputs Formation terrace, the Member B correlate, is carved in Dwyka sediments on both sides of the river (Fig. 5-32). On the south (left) bank it occurs as a narrow (less than 200 m wide), gravel and calcrete-coated bench at Douglas, and further downstream it is represented by an unimpressive scarp at Bucklands, where it also bears thin, capping gravels and calcrete. On the north bank opposite Douglas it again occurs as a bench with a thin coating of calcrete gravels. This bench continues downstream in front of the right-bank, tufa terrace at Blaauwbosdrift and Mazelsfontein. The best exposure of the thin sediments capping this terrace is at Douglas, where the bench is developed at +24 m.

<u>DOUGLAS</u>

Unit 1. "Silty Sands", 1.1 m cyclically-deposited, moderately cemented, silty sands. There are two characteristic sediments, each occurring in 5 cm thick, horizontally stratified layers with dispersed granules and fine pebbles. The first is light gray (10 YR), silty sand (no. 236). The second sediment is white (7.5 YR), silty sand (no. 237).

-- abrupt, smooth, conformable contact --

Unit 2. "Gravels", 1.1 m horizontally bedded, granule to fine cobble conglomerate (see gravel analysis in Appendix B) with a significant number of "indurated shale" clasts. The matrix of the upper 90 cm is weakly cemented, reddish yellow (7.5 YR), inhomogeneous sand (no. 235). The lower 20 cm of the gravel matrix is strongly-cemented, inhomogeneous sand (no. 234).

-- gradual, irregular, unconformable contact --

Unit 3. "Bedrock", dark blue, Dwyka shale with the upper 10 cm cemented, like the base of the capping gravel. Presumably this calcretion post-dates the deposition of unit 1.

This fragmentary deposit provides only limited information concerning river behavior during accumulation of the Rietputs-B terrace. The basal gravels indicate that the Vaal was capable of mobilizing coarse sediments, and the capping silty sands probably identify the coincident sediments on the channel margin. Post-depositional calcretion parallels cementation on this same terrace at other sites.[26]

The dominant Riverton terrace near Douglas is referable to Member IV. These brown, sandy silts and silty sands (at +10 to 12 m) occupy the apparent "floodplain" and are extensively irrigated today. Calcreted sediments referable to Member III of the Riverton Formation are found at depth as are fragments of Members I and II. All are best exposed in tributary gullies, rather than on along the river. But there are only subtle slope inflections, suggesting an early Riverton Formation terrace at about +15 m.

---

[26]For a discussion of the terraces at the Orange-Vaal confluence, see Chapter Eight below.

The events recorded by the Rietputs Formation terraces near Douglas duplicate the informative chronology established at the Vaal-Harts confluence.

Chronology of the Rietputs Formation near Douglas

1) Erosion following the abandonment of the Wedburg Terrace. River levels: near +50 m to about +27 m (at Selene).
2) Aggradation of tufas and rudaceous Vaal sediments in the highest Rietputs Formation terrace. Significant lateral erosion of Dwyka bedrocks to establish the pediment-form plains east of the river. River levels: about +27 m to near +37 m.
3) Floodplain incision and abandonment of the Rietputs-A terrace. River levels: near +37 m to less than +22 m.
4) Limited aggradation and establishment of the Rietputs-B terrace. River levels: less than +22 m to about +24 m.
5) Downcutting and abandonment of the Rietputs-B terrace. River levels: near +24 m to near +3 m.
6) Several aggradation and incision cycles of the Riverton Formation. River levels: variable, between about +3 m to above +15 m to near present.

Summary

By comparison of the Rietputs Formation sedimentation and terraces at Windsorton, Winter's Rush, the Vaal-Harts confluence, and near Douglas, it is readily apparent that a regional, systematic, geomorphic history describes the development of the modern river valleys along the lower Vaal River. The environmental history recorded by this morpho- and litho-stratigraphic sequence is examined in Chapter 10 below. But first the parallel record on the adjoining Harts, Riet, and Orange Rivers must be described.

CHAPTER 6

TERRACES ALONG THE LOWER HARTS RIVER

Label it "Qal" and color it yellow.

--Anonymous

Introduction

The Harts is a poor excuse for a river. The discharge is less than 2% of that of the Vaal (see Table 2-5), and the Harts channel, even along its lowest reach, is seldom more than 15 m wide and 2 m deep (see Figs. 6-1, 6-2, 6-3, 2-9, 5-25, 5-29). Despite its insignificant size, however, the Harts is centered in the 20 to 40 km-wide, Paleozoic trough of the Kaap Valley (du Toit, 1910; Chap. 3 above), that was reexcavated during the later Cenozoic. On the average only a few tens of meters of Dwyka argillites still line the valley perimeter.[1] Late-Cenozoic erosion has also re-established the ancient valley margins--the Gaap Escarpment to the west, and a ridge of Ventersdorp volcanics forming the Vaal-Harts interfluve on the east.

In this investigation the Harts River terraces were examined between the site of the Spitskop Dam and the Vaal confluence--a distance of 40 valley km, but over 70 river km. In this reach, as

---

[1]As noted in Chapter Three the Paleozoic perimeter of the Kaap Valley is locally "overdeepened." A bore hole at Spitskop passed through 85 m of shale before volcanics of the Ventersdorp System were intersected.

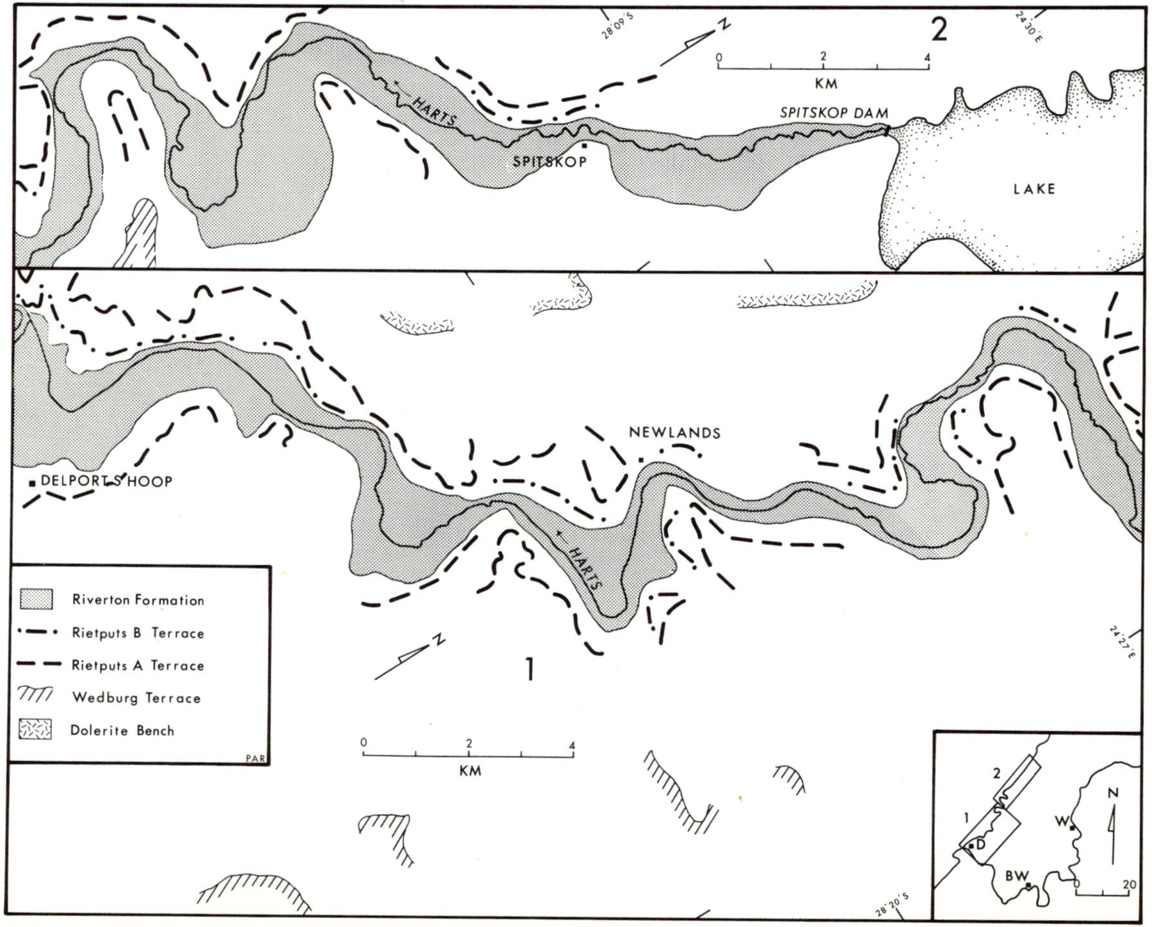

Fig. 6-1. Terraces along the lower Harts River.

Fig. 6-2. Harts River at Spitskop.

Fig. 6-3. Harts River at Newlands.

elsewhere in the Lower Vaal Basin, the modern floodplain is composed of fine-grained fills comparable to the Riverton Formation and situated below high terraces of Rietputs Formation age (see footnote 1, Chap. 5). Higher on the valley shoulders there are fragments of the Older Gravels suite. Close correlations with the alluvial history of the Vaal are indicated.

Although the Harts River rises on the subhumid plains of the southwestern Transvaal (near Lichtenburg, elevation 1480 m), the channel remains unimpressive throughout its length, even after joining the Kaap Valley and its ephemeral tributary, the Dry Harts, at Taung (elevation 1090 m). Discharge in the Harts appears to respond primarily to local rainfall and groundwater discharge from the immediate valley perimeters. Along the lower Harts the primary water source from the west is karstic discharge from the dolomites and limestones of the Gaap Plateau, while from the east the key source is runoff from the Vaal-Harts interfluve and the dolerite koppie landscape east of Spitskop. Both the ancient and modern alluvia are dominated by locally derived debris.

Older Gravels Remnants Along the Lower Harts River

The soft shales lining the Kaap Valley have a low potential for preservation of Older Gravels terraces and mere fragments of the Wedburg Terrace occur along the eastern side of the valley. In addition, potential Older Gravels sites along the upper, eastern flanks of the valley, are blanketed by red, sandy colluvia derived from weathering and erosion of andesites on the interfluve. Similarly obscured are the western valley margins, which have been coated with limestone tufas.

The oldest alluvial deposits referable to a proto-Harts River are the gravels on the Vaal-Harts interfluve at Holpan (see Chap. 4 above).[2] While these deposits have traditionally been included with the Vaal River gravels, they are equally pertinent to the early history of the Harts. Consequently the +60 m gravels along the Vaal near Windsorton also are the +150 m gravels of the Harts.

The Wedburg Terrace occurs discontinuously on the gentle, eastern slopes of the valley (see Fig. 6-1). With the exception of the prominent, calcrete veneered, shale bench south of Spitskop, these fragmentary residuals are represented by poorly demarcated, low cuestas of calcreted Dwyka shale, standing less than 5 m above the surrounding plains.[3] These features are not authentic, fluvial benches, but only escarpments that have retreated upslope since the floodplain incision at the beginning of Younger Gravels-times. These features have no associated gravels or other fluvial deposits and now stand between 30 and 40 m above the river, with the downstream examples being highest.

Also present in the Harts Valley is a discontinuous, structural bench. Exposed during the late Cenozoic, this bench is west of the river at +30 to +35 m and gives the (incorrect) impression that it is paired with Wedburg Terrace remnants across the valley. It is, however, formed by a dolerite sill subsequently coated with tufas. The possible influence of this sill on local base-levels during valley evolution is obscure.

---

[2]Older Gravels deposits and possibly planation surfaces also are present in the upper Harts basin (Williams, 1932; du Toit, 1951; Mayer, 1973). Some of these features may be relicts of ancient Vaal and Molopo drainages.

[3]In the valleys between and behind the dolerite koppies east of Spitskop (near Majeng, Seskoppies, and the Doringkloof Mine) several calcreted shale platforms persist. These features may be simply bouyed by dolerite or kimberlite sills, however, they may record fragments of low-angle valleysides of Wedburg Terrace age.

In summary, features of Older Gravels antiquity do occur in the lower Harts Valley, but they provide little information concerning ancient fluvial environments and no more than a general correlation with events along the Vaal.

Younger Gravels Terraces Along The Lower Harts River

The Younger Gravels terraces in the Harts valley generally replicate the sequence of the Rietputs Formation at the Vaal-Harts confluence, to show that the Harts and the Vaal responded to comparable environmental fluctuations during this interval. The highest terrace, correlated with Rietputs Member A, is a floodplain fill of arenaceous, Harts sediments which interdigitate with tufas derived from springs along the Gaap escarpment. The Rietputs-B equivalent terrace is only a minor, erosional bench cut into the higher terrace, while the Riverton Formation forms up to four separate terraces at even lower levels. In addition, the terraces decrease in relative elevation with distance from the Vaal confluence. For example at Delportshoop, Rietputs A is at +26 to 27 m, Rietputs B at +18 to 19 m, Riverton Member III at +14 to 16 m, while Member IV is at +9 to 10 m. At Newlands (Figs. 6-1, 6-4) these features are at +20 to 21 m, +14 to 15 m, +6 to 7 m, and +4 to 5 m respectively. And at Spitskop these features occur at +11 to 12 m, +8 to 9 m, +4 m, and +2 to 3 m. The Harts may have responded both to the fixed upstream base level provided by the dolerite sill near the Spitskop Dam and to an episodically lowered base level at the Vaal confluence.

Unfortunately there are no rudaceous sediments in this sector of the Harts valley,[4] so that there has been no diamond digging to

[4]Diamondiferous gravels which apparently are of Rietputs Formation age also occur along the Harts River between Schweizer-Reneke and Taung. These deposits were not studied.

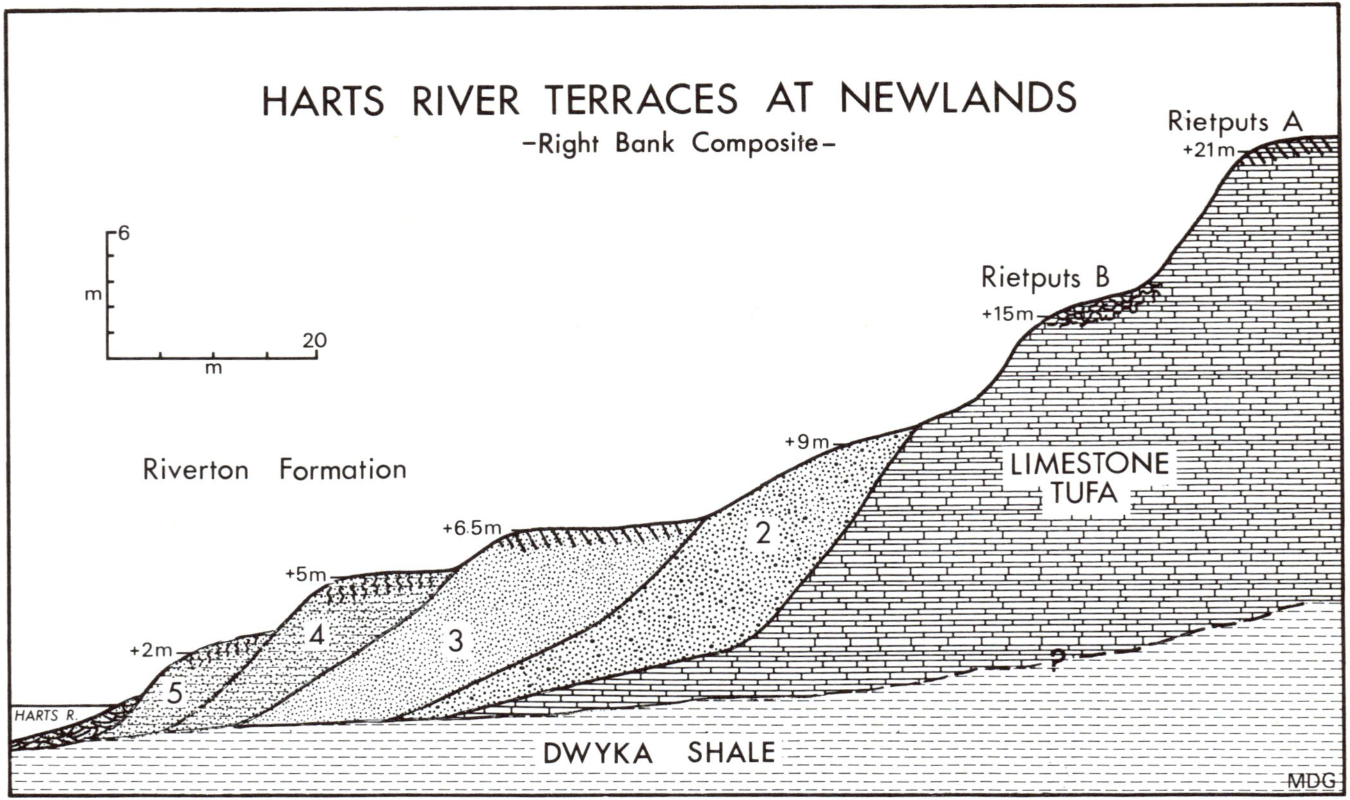

Fig. 6-4. Harts River terraces at Newlands.

produce comprehensive and informative exposures. Only shallow, surface outcrops are available. A Rietputs-A equivalent terrace is preserved subcontinuously on both sides of the river (Fig. 6-1), where it demarcates the general level to which the shale and tufa plains are graded. Along the east bank, the Rietputs-A terrace is an erosional feature developed in shale with a calcrete surface. In some areas the surface calcrete was later eroded and no longer preserves a distinct cuesta. Along the west bank, the Rietputs-A terrace is a massive valley fill of tufa and Harts alluvium. An alluvial facies of this fill near Spitskop consists of white (10 YR) calcareous sandstone which, after HCl treatment, is very well sorted, fine to medium sand (no. 204). A tufa facies further away from the modern river at Spitskop is predictably finer-grained and contains more calcite and can be described as a white (10 YR), arenaceous limestone (no. 205). No distinctive, macro-stratification is preserved at either site. The surface of the Rietputs-A tufa fill contains broad, circular solution cavities (over 100 m across, 1 to 3 m deep), filled with later colluvia that have been calcreted. Air photo examination suggests unexposed scroll topographies and related deposits are preserved along the inside of major bends, where the Harts lengthened its channel during aggradation of Rietputs Member A.

The Rietputs-B equivalent terrace typically is a calcreted, erosional shoulder in the scarp of the Rietputs-A terrace. Such features are seldom preserved (Fig. 6-1). No alluvial sediments were observed on the Rietputs-B terrace and none may exist.

The Riverton Formation equivalents along the lower Harts River are similar to the Harts River facies described at the Vaal-Harts confluence (see Chap. 5 above). Member I is dominated by pale brown to pink, colluvial and alluvial sands, Members II and III by reddish silts and sands that are usually calcreted, while Members IV and V are unconsolidated, brown silts adjacent to the modern channel (see also Butzer, 1974a; and Butzer, et al., 1978). A generalized cross-section of the Rietputs and Riverton Formation terraces at Newlands is presented in Figure 6-4.

CHAPTER 7

TERRACES ALONG THE LOWER RIET RIVER

> . . . the surface of the country is so much covered with red sand (with calcareous tufa cropping up here and there through it) as to preclude a positive statement being made with regard to the rocks beneath.
>
> -- G. W. Stow (1874: 591) in the lower Riet River valley.

Introduction

The Riet River system drains the semiarid to subhumid plains of the central and southern Orange Free State, a region dominated by near-horizontal Karoo sedimentary rocks and dispersed, intrusive volcanics. Near the Cape Province border the Riet joins its primary tributary, the Modder River, and drops through a pre-Karoo gorge to become a minor stream seldom more than 15 m wide which, like the Harts River,[1] receives only ephemeral tributaries.

---

[1] The modern terrains of the Vaal River gorge below Barkly West and the Riet River gorge are extraordinarily similar, and provide a convincing demonstration of geomorphic parallelism in landscapes that share similar lithologies and environmental histories.

During this investigation the Riet River terraces were examined upstream to the Modder confluence. Both above and below the pre-Karoo gorge (between Schutsekama and Modderrivier) the Riet valley is bounded by calcreted, shale-plains fringed by clusters of dolerite koppies and occasional outcrops of the Pre-Karoo Surface (see Figs. 2-6, 2-11).[2] The modern Riet is incised below the plains in a relatively narrow valley, containing a complex cut-and-fill sequence similar to the Riverton Formation along the Vaal (Figs. 7-1, 7-2). Terraces equivalent to those of the Rietputs Formation are found upslope at the lip of the pediment-form plains. With the exceptions of the pre-Karoo gorge upstream of Schutsekama and of the area near Plooysburg, where the Riet passes through a cluster of dolerite koppies, Rietputs terraces are nearly continuous along the lower reach of the Riet River (see Figs. 7-3, 7-4).[3]

No distinctive terraces of Older Gravels type are exposed along the lower Riet. A Wedburg Terrace is present at the Vaal-Riet confluence, but further upstream the wide, shale plains are frequently blanketed with one to five meters of red, Hutton Sands, which obscure the calcreted, bedrock profiles. Possible Riet River

---

[2]Below the Riet River gorge rocks of the Ventersdorp System remain very near the base of the modern Riet channelway and outcrop repeatedly. The largest exposure of the Pre-Karoo Surface in this reach is at the Driekops Eiland rock-engraving site.

[3]Accurate identification of fluvial terraces along the lower Riet River is complicated by several sills of Karoo dolerite. In particular, at Koedoesbergdrif, there are surfaces supported by sills at +33 to 35 m and at +52 to 56 m. Similarly, west of Plooysburg, there are surfaces supported by sills at +24 to 27 m and +30 to 33 m. While these features probably are lithologically controlled benches, they could equally be lithologically preserved, fluvial terraces. The significance of these sills in development of the Rietputs Formation terraces is a separate, but unresolved problem.

Fig. 7-1. Riet River channelway at Driekops Eiland. Terrace above wier is of lower Riverton Formation age.

Fig. 7-2. Riet River channelway at Schutsekama. Gravels in foreground may be equivalent to Rietputs C; terrace in background is of lower Riverton Formation age.

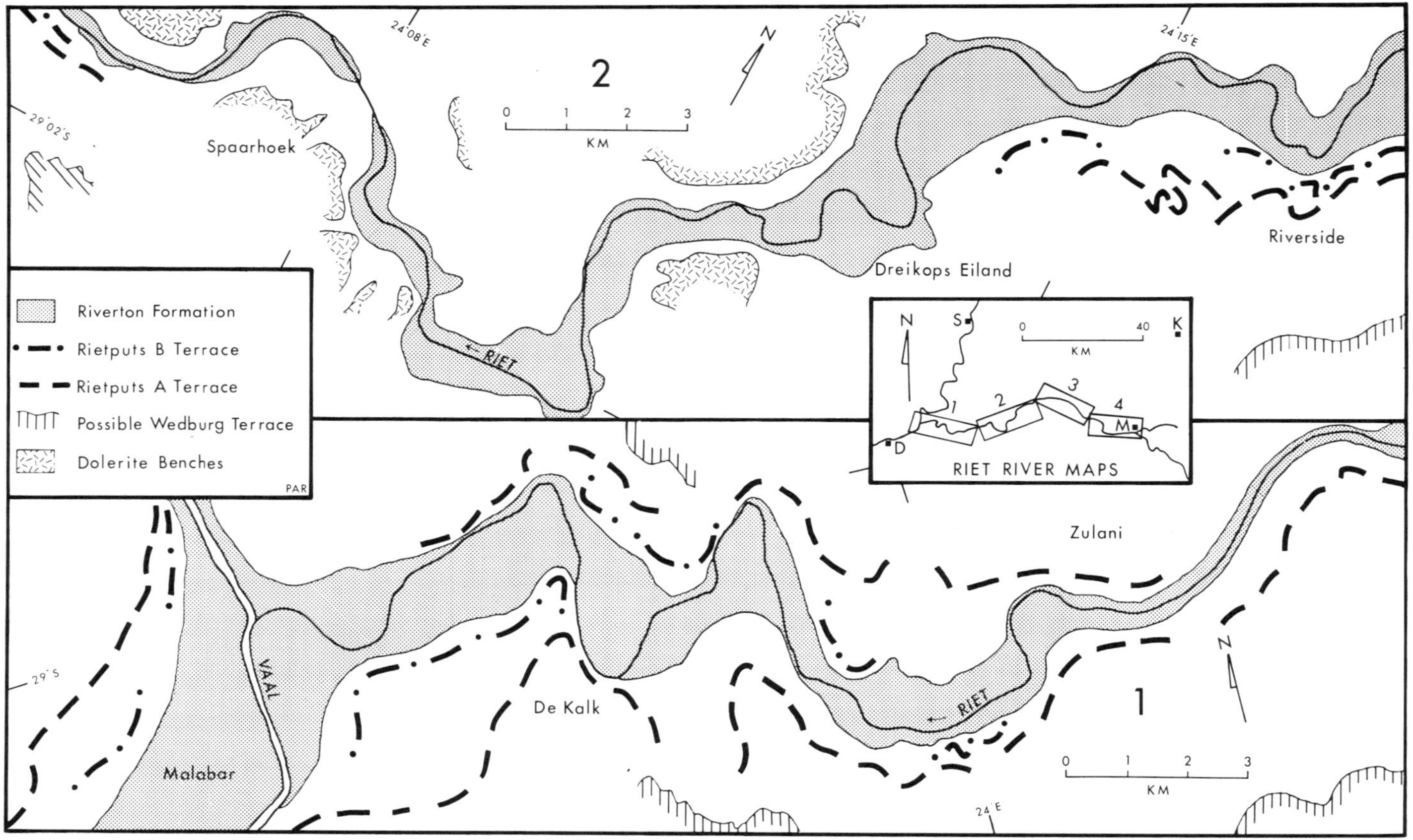

Fig. 7-3. Terraces along the Riet River from the Vaal confluence to Riverside.

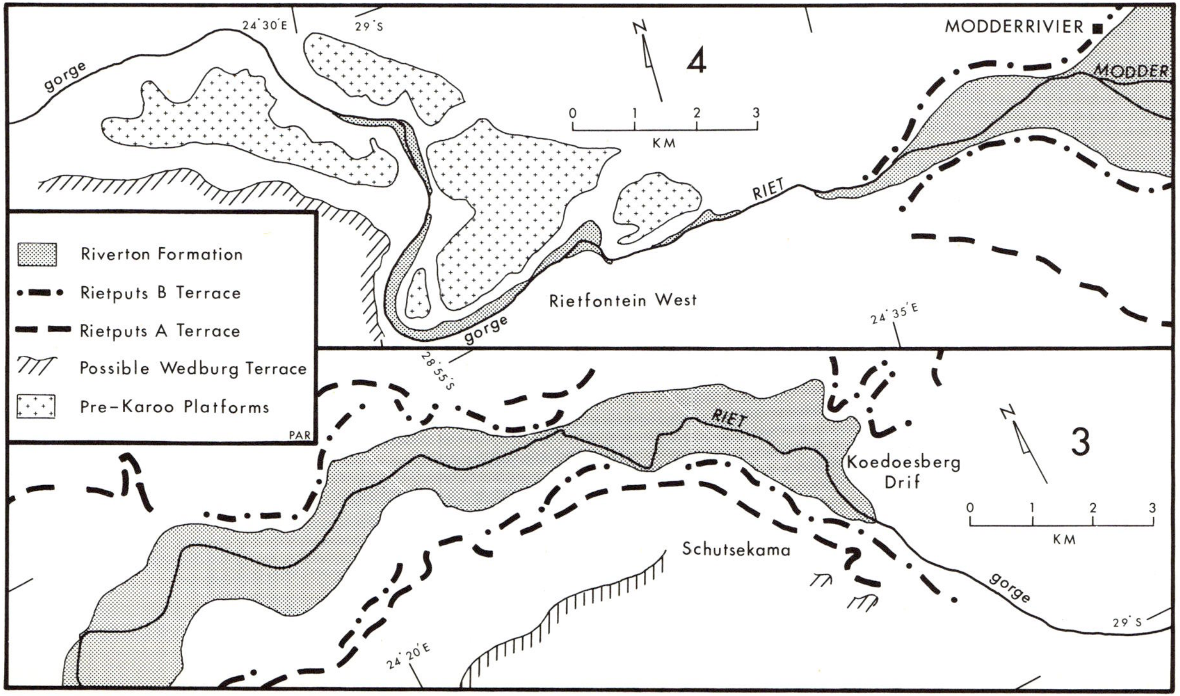

Fig. 7-4. Terraces along the Riet River from Riverside to Modderrivier.

equivalents to the Wedburg Terrace are necessarily fragmentary and limited to eroded benches, that have retreated 2 to 5 km from the valley margins and now form little more than minor inflections in shale bedrock. Sites of such features are marked on Figures 7-3 and 7-4. The minimal preservation of Older Gravels terraces in the less resistant bedrocks along the Riet River provides an additional measure for the great antiquity of the Older Gravels along the Vaal.

The Rietputs Formation and Terraces along the Lower Riet River

Rietputs Formation features of the Vaal and Riet rivers are related by morpho- and litho-stratigraphic correlations near their modern confluence (Fig. 7-3).[4] The Rietputs-A terrace is developed at +24 to 26 m all around the Riet-Vaal confluence, and consists primarily of calcreted shale or dropstone beds. Yet nearly 10 m of sediments are preserved at the top of an undercut scarp at De Kalk (Fig. 7-5); these sediments are coeval with unexposed rudaceous deposits of Rietputs A, preserved across the Vaal on Malabar. The upper 3 m of the sediments at De Kalk have low-angle, trough cross-bedding and consist of white and very pale brown (10 YR, 2.5 Y), variably cemented, silty sands that include important lenticles of fine to medium gravels (nos. 185, 186, 187). The De Kalk exposure is capped by a meter of post-depositional calcrete (no. 188). These sediments are quite different from the fine, sandy silts in the modern Riet channel. Erosional terraces referable to Rietputs B are also well developed at +16 to 18 m around the Riet-Vaal confluence, but sediments are absent from these calcreted shale surfaces.

---

[4]At the Riet-Vaal confluence the Riet River terraces descend steeply toward the Vaal. This is partly related to a dolerite dike which crosses the Riet channel above De Kalk.

Fig. 7-5. Rietputs-A terrace at De Kalk. Cliff in left foreground is 2.2 m.

The most prominent Riverton Formation feature at Riet-Vaal confluence is a +9 to 11 m terrace belonging to Member IV, while minor shoulders at +12 to 14 m suggest remnants of Member III.

Continuing upstream toward Plooysburg, preservation of the Rietputs equivalent terraces declines. Along the south margin of the valley the calcreted, shale bench of Rietputs A is subcontinuous, but along the north bank thick veneers of Hutton Sands obscure the bedrock profile and exposures of the Rietputs-A surface are sporadic. In this sector the calcreted bench of Rietputs B is rarely preserved, and forms little more than an inflection on the slope below the Rietputs-A bench. Beginning at Zulani and continuing to above Driekops Eiland, multiple dolerite sills become prominent landscape elements and few typical Rietputs equivalent terraces are preserved. These koppie terrains provide steep slopes, relatively recent, footslope veneers and many ephemeral channelways to suggest continuing dynamic hillslope sculpture. Fine-grained alluvial fills equivalent to the Riverton Formation are continuous along the modern Riet channel, and locally rest on remnants of river floor conglomerates, with Acheulian artifacts, that recall Rietputs C of the Vaal. At Driekops Eiland the Riverton Member III terrace is at +14 to 15 m, the Member IV terrace at +12 to 14 m; and the Member V terrace is at +4 m.[5]

Between Plooysburg and Schutsekama the Younger Gravels terraces reappear with a close morphologic parallel to the Riet-Vaal confluence (see Figs. 7-3, 7-4). The broad, calcreted Rietputs-A bench marks the termination of the shale plains, and is relatively continuous except where buried by Hutton Sands along the north side of the valley.[6] The Rietputs-B terrace is a narrow, discontinuous bench 5 to 8 m lower. An extensive suite of Riet River alluvia and

---

[5]Coarse gravel lenses and fills occur within the alluvial fills along the Riet River. In many cases these features simply record the erosion and reworking of clasts from a local tillite or dropstone bed. In other cases, eroded, rudaceous fills of Rietputs-B age are suggested, as are gravel lags that provide possible equivalents for Rietputs Member C along the Vaal. Clarification of the Riverton Formation equivalents along the Riet River are forthcoming (K. W. Butzer, in preparation; Butzer, Fock, Scott, and Stuckenrath, 1979).

[6]The Rietputs-A terrace residuals along the north side of the Riet valley have retreated further than the terraces on the south, and consequently occur at more irregular elevations.

terraces is preserved on Riverside, where the two Rietputs terraces are sharply defined. The Rietputs-A terrace is at +27 to 30 m, and associated sediments consist of at least a meter of calcreted, trough cross-bedded, pale yellow (2.5 YR) sands, which include dispersed fine and medium pebbles. The Rietputs-B terrace is at +22 to 24 m, with the associated sediments consisting of at least 2 m of inhomogeneous, pale yellow (2.5 Y) sands with dispersed, fine to medium pebbles and crude, horizontal stratification. Unfortunately both Rietputs equivalent deposits have minimal lateral preservation, and probably record channel-margin or tributary-channel fills. The Riverton Formation outcrops at Riverside are particularly impressive.

The Riet River gorge opens at Schutsekama, where the only thick gravels along the lower Riet are found (Fig. 7-4).[7] Episodic diamond digging has cratered the Rietputs terraces, leaving behind fragmentary exposures of the typical, calcreted, coarse gravels in inhomogeneous sandy matrices. In addition, patterns of unconformities and facies are particularly complex in alluvial fills equivalent to the Riverton Formation. The modern Riet dips steeply downvalley, as do the terraces and relict alluvia. Bedrock features along the valley at Schutsekama include (1) thin volcanic and metasedimentary beds in the Ventersdorp System, (2) planar as well as steeply dipping, exhumed Paleozoic slopes, (3) thin-bedded Dwyka tillites, shales, dropstone-beds, and limestones, and (4) multiple dolerite sills (up to 4 m thick) that are discontinuous in exposure. The highest calcreted gravels (at +36 to 38 m) are situated on a

---

[7]The geomorphic context of Schutsekama is essentially the same as that of the Vaal valley at Windsorton and Gong-Gong. Nearly all the gravel clasts are derived from tillites that once filled the pre-Karoo channelway upstream. Interestingly, there is a trace of "banded ironstone" from the Asbestos Mountains, providing additional evidence of Paleozoic glacial flows from the northwest.

gentle ridge above above the Rietputs-A terrace (at +29 to 32 m). These highest gravels may possibly be a residual of the Wedburg Terrace, although a fragmentary exposure with bedding dips of 4° to 6° suggests a colluvial residual grading down into the Rietputs-A terrace. Exposures in the latter preserve more than 3 m of massive, calcreted gravel, with discontinuous muddy and sandy interbeds. The Rietputs-B terrace is present at +17 to 20 m, and exposures along the south side of the valley suggest the same range of sediments as Rietputs A. Along the north side of the valley the surface of the Rietputs-B terrace implies a massive fill of calcreted sands, subsequently pitted by solution. Also present at Schutsekama are exposures of colluvial gravels connecting the Rietputs-B terraces, and the Rietputs-B terrace with conglomerates at base of the Riverton Formation along the channelway. The major Riverton terraces at Schutsekama are at +9 meters and +14 to 16 m.

The two Rietputs terraces persist above the pre-Karoo gorge near the Modder River confluence (Fig. 7-4).[8] Both are calcreted benches and no coeval alluvial sediments were discovered. The lower bench is at +8 to 10 m, while the higher one is at +17 to 19 m. Both these terraces replicate the rapid dip of the modern river toward the pre-Karoo gorge.

Summary

The Rietputs Formation equivalents along the lower Riet River generally are little more than sediment-free, calcreted benches standing +5 to +15 m above more recent alluvial fills of Riverton Formation age. However, these terraces provide a direct morphostratigraphic correlation to the Rietputs terraces along the Vaal

---

[8]There are no Rietputs equivalent terraces in the Riet River gorge. However, there are a series of equivocal slope inflections and possibly equivalent talus screes on Knoffelfontein and Riet Fontein West.

and Harts Rivers and, together with the rare exposures of comparable sediments, demonstrate that the record of the Rietputs Formation is of regionwide significance, and ultimately related to regional patterns of formative, geomorphic environments.

CHAPTER 8

TERRACES ALONG THE MIDDLE ORANGE RIVER

> The country further down the Orange River is formed in great part by the Dwyka, and the numerous boulders weathered out of the latter, together with the dense growth of thorntrees, chiefly the unpleasant 'hakdoorn' render geological exploration difficult.
>
> -- A.L. du Toit (1908: 174)

Setting of the Orange River

The Orange is the great river of the South African interior. Together with its other principal tributary, the Caledon, it rises above the 3000 m contour, on the west slopes of the Drakensberg, to drain much of Lesotho and the adjoining plains of the Orange Free State and northeastern Cape Province, before reaching the Vaal. This highland source usually provides the Orange with an impressive perennial flow, on which late-summer floods are superimposed (Fig. 1-1 and Table 2-5). Below the Caledon confluence, the Orange receives only minor, ephemeral tributaries, but even the Vaal discharge is dwarfed by the Orange (see Fig. 8-1). Apart from local gorges of pre-Karoo rocks, the upper Orange basin consists of near-horizontal Karoo System sedimentaries and volcanics.

Fig. 8-1. The Orange-Vaal confluence, March 1973 (Vaal at left).

In this investigation the Orange River terraces were examined from Hopetown, past the Vaal confluence to the foot of the Gaap escarpment (approximately 125 river km, Figs. 8-2, 8-3, 8-4). The landscape north of this river sector is dominated by irregular ridges of Ventersdorp System rocks (the Spioenkop Anticline, see Chap. 3 above), while to the south of the river there are extensive plains of calcreted, Dwyka sedimentaries (see Fig. 3-1). At Hopetown the Orange (at 1052 m) enters a gorge cut into pre-Karoo (Ventersdorp) rocks, which opens 54 river km downstream at Torquay. Along this reach the Orange decends over 60 m. While this gorge definitely demarcates a chain of small, pre-Karoo valleys still partly filled with Dwyka sediments, it was not possible to verify this feature as a pre-Karoo channelway. The Orange, with a valley size comparable to the Vaal, but with annual flows 2 to 25 times greater, has had

Fig. 8-2. Orange River valley below Read's Drift.

much less difficulty in eroding resistant bedrock thresholds of the Pre-Karoo Surface (Fig. 2-10). Below Torquay, the Orange traces a pre-Karoo channel only incompletely exhumed, where pre-Karoo rocks are initially only present on the floodplain margins below higher ridges of calcreted Dwyka sediments. But below the Vaal confluence a deep, pre-Dwyka trough funnels the combined flow out of the region. In the 72 river km between Torquay and the end of the study area at Read's Drift, the Orange drops a further 38 m.

Terraces equivalent to the Older Gravels and Rietputs Formation along the Vaal also are found along the Orange, continuing the theme of episodic, later-Cenozoic exhumation of Paleozoic valleys. Unfortunately the coincident, alluvial sediments are rarely preserved and correlations can only be established on morpho-stratigraphic grounds. However, aggradational surfaces along the modern Orange

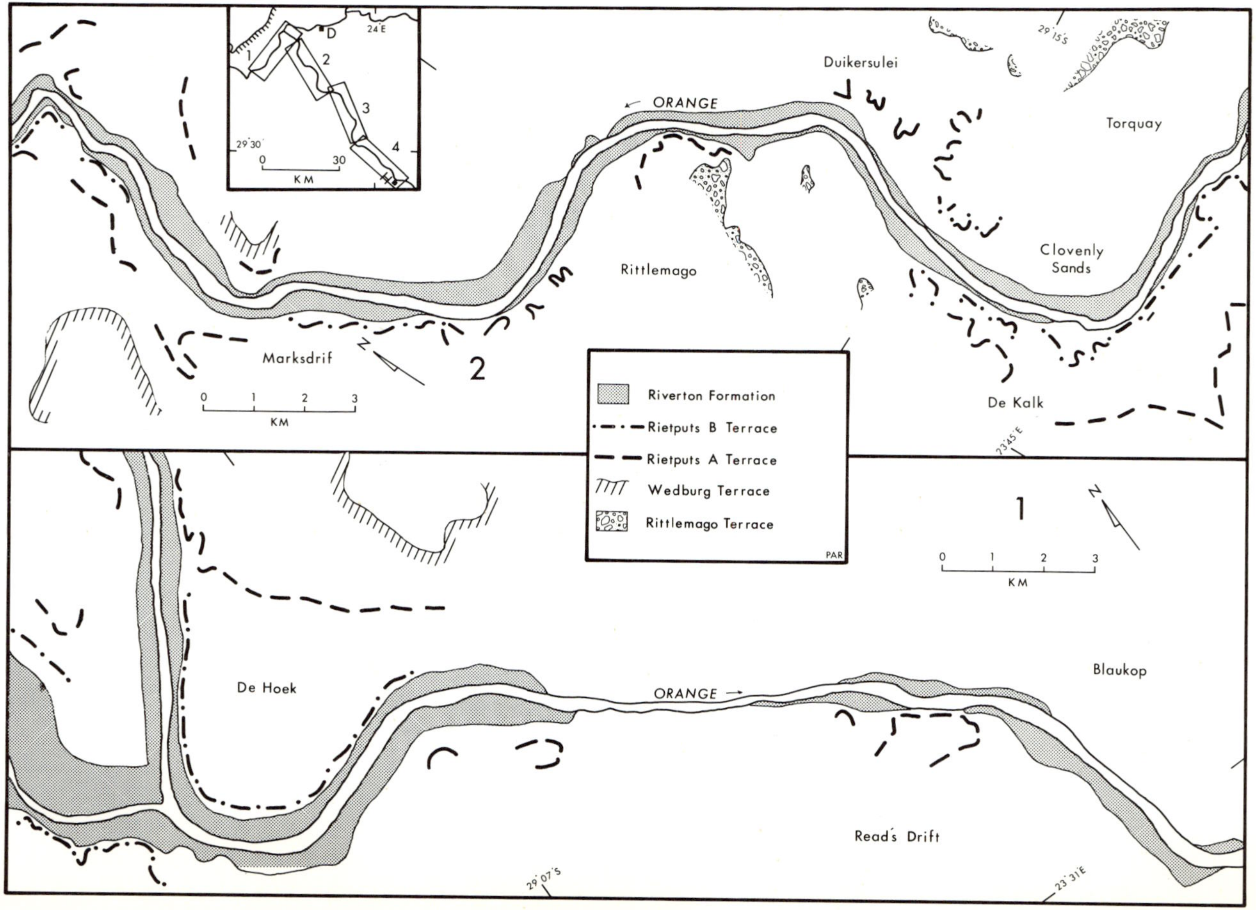

Fig. 8-3. Terraces along the Orange River, Read's Drift to Torquay.

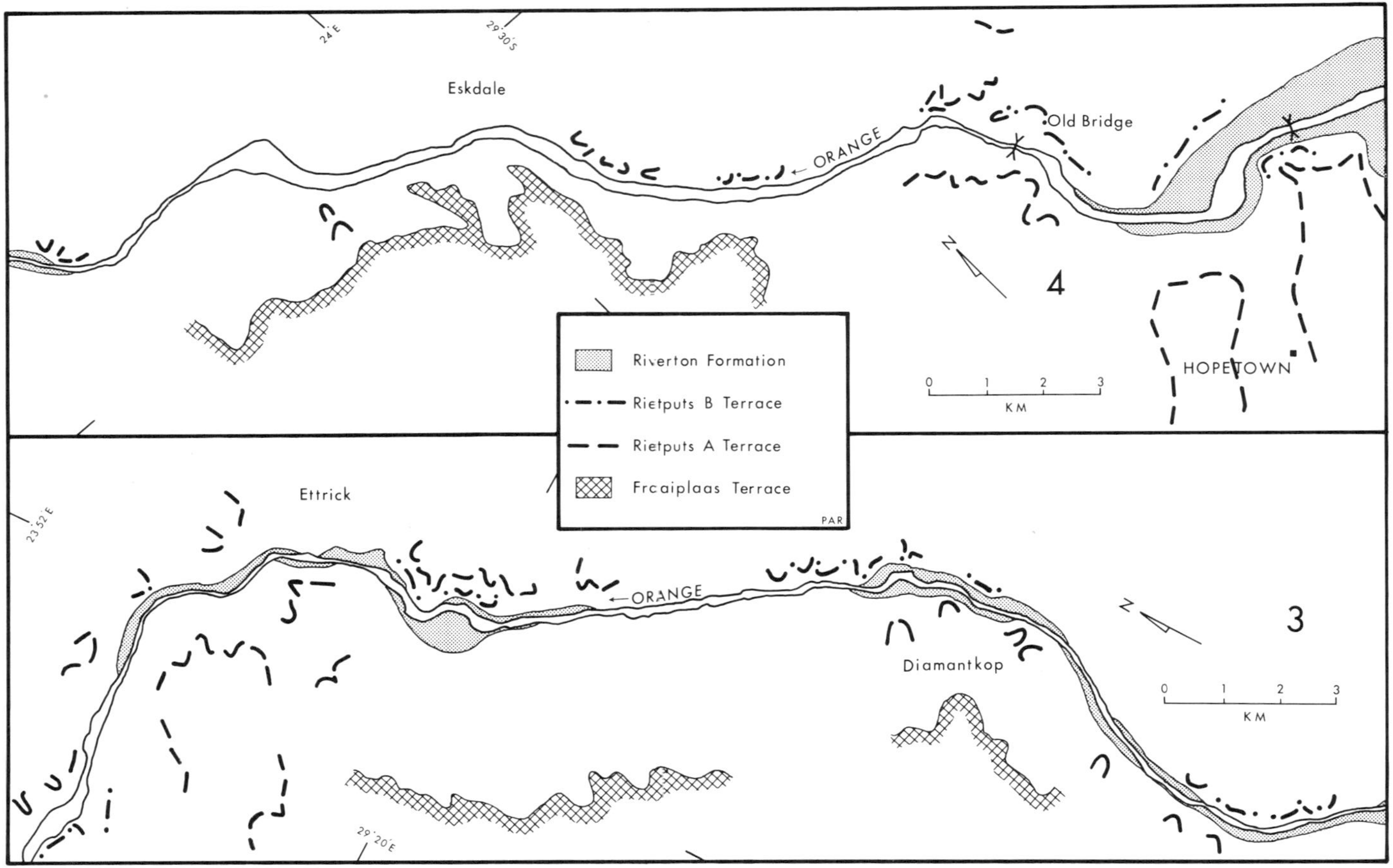

Fig. 8-4. Terraces along the Orange River, Klipfontein to Hopetown.

channel are composed of multiple alluvial fills more recent than the Rietputs Formation, providing direct parallels with the Riverton Formation.

Older Gravels Terraces Along The Orange River

There are three Older Gravels terrace complexes along the middle Orange River. The lowest of these fits perfectly with the profile of the Wedburg Terrace along the lower Vaal. The highest Older Gravel feature indicates an Orange channelway far south of the present axis, while an intermediate terrace trace appears to record a transitional stage.

The highest Older Gravel channel is recorded by calcreted escarpments of Dwyka sediments that extend along the southern margin of the valley from near Hopetown to Klipfontein[1] and Fraaiplaas (Figs. 8-5, 8-6, 8-7). Eventually this channel trace rejoins the route of the modern Orange at the high-level alluvial gravels on Kransfontein and Lanyon Vale (see Fig. 8-5). This feature is here named the Fraaiplaas Terrace.

Between Hopetown and Klipfontein, the base of the Fraaiplaas scarp begins Ventersdorp rocks behind the modern Orange floodplain, but extends upslope through Dwyka tillites and dropstone beds before its calcreted summit is reached. However, the summit surface dips in this sector away from the modern river for a distance of several hundred meters before continuing its general rise to the south. This reversed dip suggests that the Fraaiplaas Terrace has retreated from its original site, no longer exactly demarcating the elevation of the former floodplain. This observation is emphasized

---

[1]Farm names are repeated frequently, even within the same magisterial district. This Klipfontein is not the famous rock engraving site, now part of the farm Rooipoort (northeast of Schmidtsdrif).

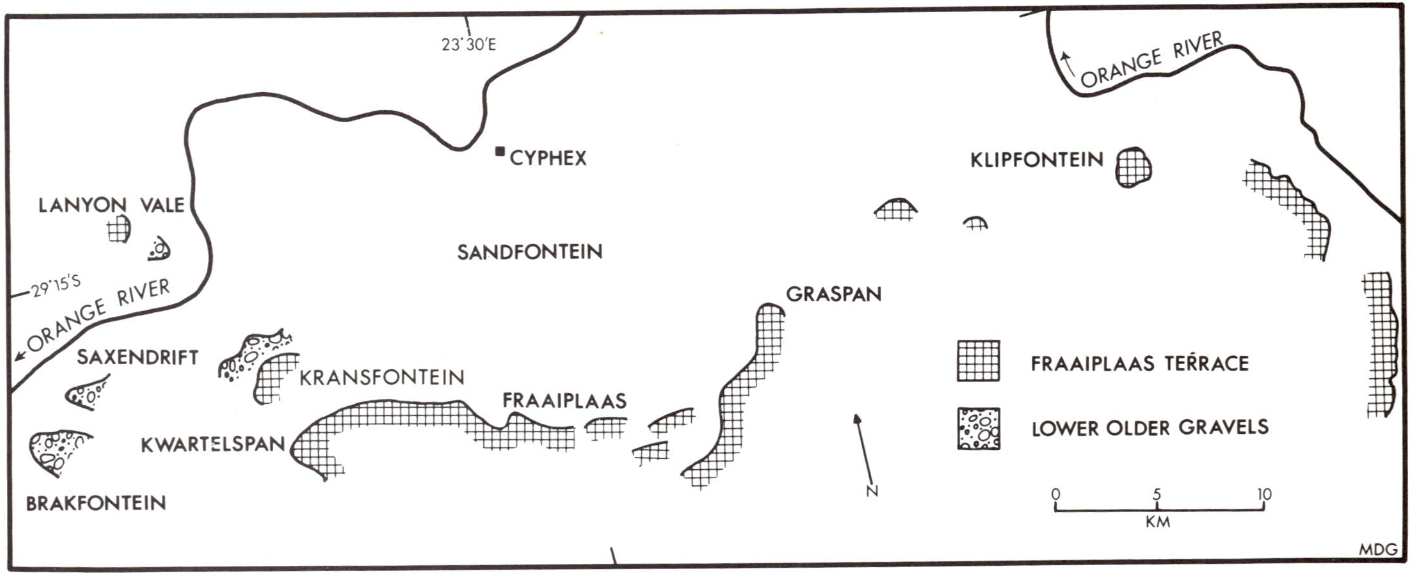

Fig. 8-5. Trace of the Fraaiplaas Terrace.

Fig. 8-6. Fraaiplaas Terrace at Fraaiplaas.

Fig. 8-7. Abandoned diggings in the Older Gravels at Klipfontein (view to the south).

by the modern elevations of the scarp crest, varying between 1090 and 115 m (+65 to 80 m above the modern Orange). The adjacent shale/tillite plains additionally appear to have been regraded since the time of the Fraaiplaas Terrace channelway. At Klipfontein calcreted surficial washes of the typical, granule-to-boulder grade gravels, presumably derived from primary, Fraaiplaas Terrace deposits, coat a knoll of volcanic rocks. Continuing on the shale plains west of Klipfontein the calcreted Fraaiplaas Terrace occurs as an isolated scarp rising to +30 m and extending almost continuously for nearly 40 km between Graspan and Kwartelspan. This scarp also has an uneven summit elevation (1060 - 1110 m) again indicating retreat from its original site. Furthermore, since no deposits diagnostic of a major river are exposed along this distance, the exact channel route cannot be demarcated. Continuing the trace of the Fraaiplaas Terrace westward, the diamond diggings along the Orange at Kransfontein are intersected, and generalized, morpho-stratigraphic correlation of these deposits with the Fraaiplaas Terrace must be assumed. Since no primary, Fraaiplaas Terrace channelways are preserved, reconstructing the coeval fluvial dynamics is not possible. It is unclear whether the Fraaiplaas Terrace corresponds with any Older Gravels features along the Vaal. However, if the Geduld Bench (see Chap. Four above) were extended downstream, it would grade to the level of the Fraaiplaas Terrace at Kwartelspan.[2]

---

[2]In terms of long-distance morpho-stratigraphic correlation, the Fraaiplaas Terrace could correspond to either the +30 meter (Proksch Koppie) or +60 meter (Holpan) gravels near Windsorton. As noted earlier, the oldest (Cenozoic) morpho-stratigraphic element in the Lower Vaal Basin is an 8 m-thick tufa that caps the southern margin of the Gaap Escarpment at Witberg.

The middle elevation, Older Gravels channel-trace along the Orange is fragmentary. It is marked by a retreating scarp of Dwyka sediments north of Torquay, by an isolated exposure of reworked gravels east of Torquay, and by an impressive, gravel-coated, shale and tillite surface at Rittlemago which dips down from this former terrace level (Fig. 8-3). The original level of the terrace at Rittlemago is no longer marked by a major platform, although an isolated, gravel-capped peak immediately east of Rittlemago occurs at an elevation of 1050 m, +70 m above the modern Orange. The reworked gravels on the low-angle surface at Rittlemago are up to 3 m thick and have crude, horizontal bedding. The gravels are composed of typical, subrounded, granule-to-coarse cobble-grade clasts of andesite, with subsidiary Ventersdorp quartzites and indurated shale (see Gravel Analysis in Appendix B). No doubt most of these clasts were originally derived from local tillites and dropstone beds.[3] Gravel matrices are a calcreted, clayey silt sand (no. 225). The most economical explanation for the Rittlemago gravels sees these fragmentary, colluvial deposits recording complex channel adjustments as the Fraaiplaas Terrace channelway was captured by a tributary of the Vaal. However, there is insufficient evidence to verify this speculation.

The lowest Older Gravels terrace along the Orange is equivalent to the Wedburg Terrace along the Vaal. It is recorded by a platform near Marksdrif at +40 m (relative to the Orange at Marksdrif, +52 m relative to the Orange below De Hoek, Fig. 8-3). Surface sediments consist of more than 2 m of planar-stratified pebbles and cobbles (well sorted, partly imbricated) set in a calcreted matrix of very pale brown (10 YR), silty sand. These textures are similar to

---

[3]As along the Vaal, most relict gravel clasts along the Orange have their ultimate sources in Dwyka bedrocks.

local gravel splays in the channel of the modern Orange. Additional Wedburg Terrace outcrops of calcreted shale are present near Duikersvlei. Since these Wedburg Terrace equivalents were only discovered near the Vaal confluence and further downstream,[4] the Wedburg Terrace may only be a feature of the Vaal drainage.

Rietputs Formation Equivalents Along The Middle Orange River

The Rietputs A and B terraces along the Vaal near Douglas continue to the Orange-Vaal confluence, providing a morpho- and litho-stratigraphic correlation for the two river systems. The Rietputs Formation terraces along the Orange are platforms cut in Dwyka sediments, that bear thin veneers of calcreted, alluvial lag deposits.[5] However, exposures are infrequent, thick gravel deposits of interest to diamond diggers are unusual, and most valley flanks are covered by modern and/or relict dune sands--the Clovenly Sands (see Chapter Two above), episodically deflated from

---

[4]High level gravels downstream at Saxendrift and Brakfontein may correlate with the Wedburg Terrace.

[5]The preservation of Rietputs Formation terraces along the Orange is strongly localized to sites where Dwyka sediments continue to fill pre-Karoo valleyways. It is appropriate to emphasize that exhumation of the pre-Karoo valleys is primarily due to the hillslope processes, not corrasion along the channel perimeter. The Rietputs terraces are largely the product of hillslope planation during intervals of channel profile stability. Since hillslope erosion is dependent on immediate rock weathering trends, processes of sediment removal, as well as slope gradient, all of which are only distantly related to the corrasive power of the stream at the valley bottom, we should expect and do find similar geomorphic themes in the exhumation of the pre-Karoo valleys along the Orange, Vaal and Riet rivers--despite the modern and historical variations in discharge between these streams.

Orange River channels during low flow seasons since early Riverton times. These sands are thickest along the eastern side of NW and SW-trending reaches.

The mask of Clovenly Sands is particularly effective inside the bend of the Orange River at the Vaal confluence, where a morphological parallel to the Vaal-Harts confluence is evident. Channel lengthening has been the local theme since abandonment of the Fraaiplaas Terrace. The calcreted summits of the Rietputs-A terrace (at +28 to 30 m) and the Rietputs-B terrace (at +20 to 22 m) protrude only discontinuously on De Hoek and Marksdrif (Fig. 8-3). At the former, rare exposures in both terraces expose gravels up to 6 m thick, but there are too few exposures to develop a systematic stratigraphy. On Marksdrif both terraces are benches cut into Dwyka tillite with a thin layer of calcreted gravels. There also is a low-level gravel deposit, below +8 m, at De Hoek that is comparable to the well-sorted, coarse pebble gravels across the Orange-Vaal confluence at Mazelsfontein. Preserved at Mazelsfontein is a +3 m platform in Dwyka shale, covered by (1) 4 m of large pebble conglomerate, (2) 2 m of calcrete, and (3) 7 m of reddish yellow, sandy silt. The well-sorted conglomerate has well-developed horizontal stratification and a brown (7.5 YR), variably-cemented matrix of inhomogeneous sands (nos. 246, 247). The white (10 YR), moderately-cemented calcrete has a residual of clayey silt sand after HCl treatment, and includes many clasts of weathered quartz (no. 249). The upper, reddish yellow, sandy silt (no. 248) is 44% HCl soluble, even though only slightly cemented. The two upper layers belong to the early Riverton Formation, and the upper surface records a +16 m terrace, 3 to 5 m above the Member IV floodplain.[6]

---

[6]Riverton Formation terraces above the dominant Member IV surface are rare along the Orange River, and the erosional phase preceding Member IV appears to have been particularly effective. Earlier Riverton fills are mainly confined to tributary gullies well above the floodplain.

The basal conglomerate may be of earliest Riverton Formation age or, more likely, an Orange River variant of Rietputs Member C.

As the Orange descends the pre-Karoo gorge below the Vaal confluence, fragments of Younger Gravel terraces are present wherever fills of Dwyka sediments are preserved. Only the Rietputs-A terraces are impressive, while Rietputs-B usually is little more than a minor shoulder, discontinuously developed along the valleyside. The largest terrace is a +40 to 43 m, calcreted bench at Bloukop, while at Read's Drift there is an outcrop of calcreted gravels atop an andesite ridge (Fig. 8-3).

Although typically discontinuous and often poorly exposed, Rietputs terraces are present along the flanks of the middle Orange River valley from Marksdrif upstream to beyond Hopetown (Figs. 8-3 and 8-4). At Duikersvlei the Member-A terrace (at about +30 m) emerges from beneath the Clovenly Sands with a typical surface of calcreted, coarse cobble gravels, about a meter thick. Member-B terraces are present immediately upstream and across the valley. The most impressive Rietputs terraces along the middle Orange are at De Kalk[7] and Klipfontein. Here both terraces appear on the left bank as major benches cut into Dwyka sediments, each bearing a thin layer of calcreted gravels. The Member-A terrace is at +40 to +42 m, while the more sharply outlined Member-B terrace is at +18 to +20 m. The equivalent sites on the right bank, at Torquay, are buried by Clovenly Sands.

In the steep, incised reach between Klipfontein and Hopetown the two Younger Gravels terraces occur sporadically (Fig. 8-4), with the east side terraces poorly exposed beneath a mantle of Clovenly Sands. Several calcreted escarpments referable to

---

[7]This farm De Kalk is not the farm by the same name along the lower Riet River. De Kalk (on Orange) is famous as the site where the first diamond was (officially) discovered in South Africa (1866).

Rietputs A appear to have retreated more than 2 km since their formation. Deposits equivalent to Rietputs A occur below Diamantkop peak at +37 to 40 m. At this site a 5 to 8 m thickness of complexly-bedded, alluvial sediments form a conformable deposit on tillite bedrock. These calcreted sediments include poorly sorted, granule-to-boulder-grade gravels with crude, horizontal bedding, well-sorted boulder-grade gravels with trough crossbedding, well-sorted coarse sands with planar crossbedding, as well as well-sorted medium to coarse pebbles with both planar and trough cross-bedding.

At Hopetown the Rietputs-A terrace is present on both sides of the river near the old bridge, but it is best developed as a +27 to 29 m bench of Dwyka tillite on both sides of the village (Fig. 8-4). The calcreted surface of this platform includes 2.5 m of poorly exposed, granule-to-coarse-cobble gravels. The Member-B terrace is also present near Hopetown, particularly on the right bank above the old (downstream) bridge (at +22 to 24 m). The major Riverton Formation terraces at Hopetown are at +7 to +8 m (Member V), +12 to +14 m (Member IV) and +16 to +17 m (Member III).

This brief summary of the terraces along the middle Orange River serves to emphasize that the terraces of the Lower Vaal River are part of a regional morpho-stratigraphic system, which extend across variable bedrocks to be nearly continuous for long distances along streams that have sources in different precipitation zones. Unfortunately only the Vaal provides morpho- and litho-stratigraphic records informative of past fluvial dynamics. These issues are considered at length in Chapter 10.

CHAPTER 9

CENOZOIC DENUDATION IN THE LOWER VAAL BASIN

So it goes.
-- Kurt Vonnegut, Slaughterhouse 5

L. C. King and the Geomorphology of South Africa

In a series of papers beginning in the 1940's and culiminating in South African Scenery (1963: 3rd ed., rev.), Morphology of the Earth (1967: 2nd ed.), and The Natal Monocline (1972), Lester C. King has proposed a comprehensive theory of landscape sculpture and denudation history to explain the geomorphology of southern Africa. With only few exceptions (e.g. de Swardt and Bennet, 1974) King's ideas have been uncritically embraced in South Africa and now are virtual dogma. In this chapter King's paradigm is analyzed briefly from both theoretical and empirical perspectives, and is found to be little use in explaining the geomorphic history of the Lower Vaal Basin. In particular, the record of denudation in the Lower Vaal Basin does not comply with King's model, and related interpretations of the later Cenozoic alluvia (e.g., Partridge and Brink, 1967) are not viable.

To explain the progressively higher planar landscapes in South Africa, King built on the arguments of William Morris Davis (1906) and Frank Dixey (1938, 1942) to postulate his "Standard Cycle of Erosion (Pediplanation)" (1967: 159-168). In briefest terms, King proposed that landscape erosion occurs universally through slope retreat (backwearing).[1] Beginning at a coast, such erosion produces a planar landsurface which ever-widens as the base of the retreating slope (scarp) moves inland. Upon uplift of the landmass, or long term marine regression, a new interval of erosion begins at the coast and a new scarp begins retreating inland. This scarp consumes the plain of the previous cycle at its seaward margin, while further inland the scarp of the previous cycle keeps widening the same plain as it continues to retreat, unaffected by the sea-level change at the coast. Land surfaces are obliterated when scarps from opposite sides of the continent meet, and consume each other. Given repeated episodic uplift over long time intervals, this model predicts progressively higher inland plateaus bounded by retreating scarps. Following this theory, the river talwegs behave much the same as the slopes. Coastal uplift produces a knickpoint in the talwegs of the streams draining to the coast and these knickpoints also retreat inland (conveniently at about the same rate as the scarps). Thus the rivers draining the landmass are largely consequent on the land surfaces, with low gradients, as they flow over the plains and knickpoints where they pass through retreating scarps. Furthermore, King postulated that these erosion cycles proceed inland with general disregard for variations both in regional climate and in the lithology and structure of the continental rocks.

After using the interior plateaus of South Africa to inspire his theory, King then turned to these same planar surfaces to demonstrate it. He explains the physiography of South Africa accordingly in The Morphology of the Earth (pp. 241-309) and South African Scenery. King postulates the geomorphology of South Africa

---

[1]For a longer review of King's ideas, see de Swardt and Bennet (1974).

in terms of four major planation intervals (or at least intervals of important valley widening) since the Triassic. The oldest and highest of these is the "Gondwana" landscape, supposedly preserved in the Lesotho highlands and isolated mountain summits in the Transvaal and Cape Province. King maintains this surface was cut across Gondwanaland during the late Triassic and Jurassic, prior to its fractionation. Following the break-up of Gondwanaland and the coeval tectonic stresses, the "African" surface was planed across the subcontinent between late Cretaceous and mid-Cenozoic (?Oligocene). According to King, this is the High Veld surface, which he correlates with the high plains of the northern Cape Province and Karoo. In the Miocene(?) another interval of planation (the "post-African") opened the river valleys (in perhaps two phases) to produce what King called the "Cape Middleveld"--the plains of the southern Karoo, Rhodesia, and the northern Transvaal. Supposedly a fourth planation interval begun during Pliocene times is demonstrated by planation surfaces along the coasts. An interval of strong valley incision concludes King's chronology, and he claims to find a demonstration for this in the falls of the lower Orange, Zambezi, and Cunene rivers. Of course, when explaining the forms of smaller regions in South African Scenery, King includes enough local lithological and structural qualifications to produce a believable story.

Some Canons of Landscape Evolution

King's theory of landscape development is dependent on three key axioms--first, that southern Africa was subject to episodic uplift during Cenozoic times; second, that all slopes not only retreat (backwear), but that they retreat inland for long distances; a third, that river knickpoints can also retreat inland for long distances. None of these propositions is supported by the available empirical or theoretical evidence.

King maintained his erosional cycles should be mirrored by depositional cycles offshore. Pulses of deposition consequent to uplift and initiation of a new erosion cycle should be interspersed with intervals of slow sedimentation. Correlating tectonic deformation with the magnitudes of clastic loads in streams remains a difficult problem, but King interpreted the coastal deposits of Natal and Moçambique accordingly. However, recent exploration of the deposits off the Orange River mouth and along the Natal coast seem to contradict this. The deposits on the continental shelf off the Orange River mouth (see Emery, 1972; Dingle, 1973) suggest decreasing but relatively continuous sedimentation since the early Cretaceous. There is a noteworthy erosional break of mid-Tertiary age and several minor depositional hiatuses. The J(c) - 1 bore hole off the Natal coast (see duToit and Leith, 1974) also records a decrease in sedimentation during the mid-Tertiary (? Oligocene), an unconformity between the Eocene and Oligocene beds, and several lesser, depositional breaks. However, strict correlation of these distant sites is not to be presumed. There are many possible explanations for the changing sedimentation rates and unconformities. The best alternative sees the sedimentation patterns as a consequence of decreasing suspended stream loads, caused by progressively more arid climates on the adjacent coastal forelands.[2] The unconformities may be the product of irregular, but progressive isostatic adjustments as the coastal shelves subsided or tilted outward. Even more poorly understood problems of global tectonics

---

[2]The latitudinal displacement of southern Africa since the Mesozoic has been northward, progressively moving the subcontinent from wetter climates dominated by the mid-latitude westerlies to the drier climates dominated by the subtropical anticyclones. Similarly the more rapid, early Cenozoic deposition of off-shore zones may have been a consequence of humid climates flushing seaward unconsolidated debris that was left in the Gondwanaland interior by relatively arid, "continental" climates of the Mesozoic.

and world ocean levels may be involved. In any case these offshore deposits do not verify episodic continental uplift since the Mesozoic. If anything, progressive, relatively continuous, continental-scale, isostatic uplift is indicated, which is an expected response to large-scale erosion of the sedimentary covers on ancient cratons (Garner, 1974: 128-130).

The processual background to King's pediplanation and slope retreat has been criticized by Frye (1958) and more recently (and severely) by Garner (1974: 15-16, 328-329). The surveys of Young (1972) and of Carson and Kirkby (1972) clearly demonstrate that not all slopes "backwear". Some slopes "backwear," some "downwear," and some do both--depending upon the character and effectiveness of the erosion processes, the modes of rock weathering, and the bedrock lithologies. While slope retreat for tens of kilometers does seem demonstrated in some settings, such as on the Colorado Plateau, the east slope of the Drakensberg, and the Khomas Hochland (South West Africa), long distance retreat is not demonstrated by any of the world's landscapes, nor is it to be expected. The lithologies and processes of erosion and weathering that produce slope retreat could hardly be expected to be continuously available across the diverse rocks and environments found on any continent.

King never fully explains how he expects stream knickpoints to retreat ever inland. The implication is that knickpoint retreat is a consequence of hillslope retreat, even though the knickpoints marginally precede the scarps as they march inland. Little, if any processual research has been directed to this problem. Obviously the erosion processes that operate in stream channels are very different from those on slopes, and morphologic symmetry is not to be expected. Many of the world's waterfalls (e.g. Victoria Falls, Augrabies Falls, Niagara Falls) demonstrate that knickpoints in selected bedrock settings can retreat for short distances, but retreats of hundreds of kilometers are not demonstrated anywhere.

The overall trend is knickpoint destruction and smoothing of the stream profile, although uniformity in detail is seldom achieved (Morisawa, 1968: 107-110). Knickpoints only seem to persist in terrains with resistant lithologies, and such lithologies are hardly continuously available on a continental traverse.

By the Evidence

As if these conceptual issues were not telling enough, ample empirical contradictions to King's theories are available within and without the Vaal basin.

(1) De Swardt and Bennet (1974) have proposed a near-complete revision for the planation surfaces in Natal which remains to be effectively answered (see King, 1976).

(2) Along most of the south coast of Cape Province an impressive planation surface with associated fanglomerates stands at about +200 m and reaches inland up to 15 km (see Butzer and Helgren, 1972). This surface is a classic expression of King's supposed Pliocene planation. However, estuarine deposits under capping eolianites have been dated palynologically as Oligocene, implying an Eocene minimum-age for erosion of this Coastal Platform (Helgren and Butzer, 1977). If King's model is pursued, then the absurd correlation of this coastal surface with the High Veld and other elements of the "African" surface would necessarily follow.

(3) In the Vaal basin the first major knickpoints below the High Veld surface occur along the Vaal River at Warrenton, and the Riet River, near the Modder River confluence. At both locations the surrounding plains "grade" to about river level, which is supported by a surface formed in resistant igneous rocks. However, the surface is exhumed, preserving markers (striations and sedimentary fills) of late Paleozoic glaciation (Chap. 3 above; see also Wellington, 1937). Such a knickpoint is hardly referable to Cenozoic denudation.

(4) The slopes on the High Veld are undulating, concavo-convex in profile. King considered them to be the products of "moribund" pediplanation. Yet it is clear that the drainage divides and the drainage lines have maintained relatively constant positions since these planation features developed, and net surface lowering must have occurred largely through slope downwearing.

(5) None of the six basic categories of planar landforms in the Lower Vaal Basin can be related to King's denudation chronology.[3] These six categories are: (a) Paleozoic, planar elements of the exhumed Pre-Karoo Surface, which include the summit of the Gaap Escarpment, the Nooitgedacht, Holpan, and Proksch Koppie Platforms along the pre-Dwyka, Vaal River, as well as the planation features along the Riet River gorge (see Chapter 3 above); (b) Structural plains supported by sills of Karoo dolerite, which are best developed as cap rocks on koppies; (c) Now-calcreted erosional plains that developed in response to local base-levels provided by the Pliocene Wedburg-stage in the Vaal River chronology (see Chap. 4 above); (d) Now-calcreted plains also developed in response to the local base-levels provided by the Pleistocene Younger Gravels in the Vaal River chronology (see Chap. 5 above); these nearly ubiquitous, early Pleistocene features merge with the Wedburg-age plains in the lower Harts and Riet valleys; (e) Mid-Pleistocene to Holocene river

---

[3]On the 1:16,000,000 map in Morphology of the Earth (1967: Fig. 119) King seems to suggest that the modern terrains of the Lower Vaal Basin above 1200 m (4000 ft.) are part of his early Cenozoic, "African Surface". These include the summits of the Gaap Escarpment and Vaal-Harts interfluve (both verified here as exhumed Paleozoic terrains) and the crest of the Riet-Orange interfluve along the Spioenkop Anticline (see Chap. 3 above), which owes its relief to the combined resistence of the Ventersdorp andesites and Karoo dolerites. King categorized the lower terrains in the region as part of his "post-African surface" (Cape Middleveld) of "late Cenozoic" age.

terraces (Younger Gravels and Riverton Formation); and (f) Modern pan basins and their relict levels recorded by dissected pan fills (Butzer, Fock, et al., 1973; Butzer, 1974b); the interfluves between pan basins include igneous thresholds and erosional bevels of Younger Gravels-age or older, while the artifacts on the pan terrains compare with those in Riverton Formation contexts.[4]

While King's explanation of the planar landscapes in South Africa is no longer satisfactory, the planar landscapes still exist and demand explanation. Explanation can be sought only through detailed regional studies. When a sufficient number of small regions have been intensively studied, and when a better understanding of the tectonic history of southern Africa has been obtained, the situation will be ripe for more ambitious synthesis.

---

[4]These empirical examples from the Lower Vaal Basin do not demonstrate that cyclic knickpoint recession did not occur. They only verify that if knickpoint retreat did occur, it was of no geomorphic importance.

CHAPTER 10

ENVIRONMENTS OF THE RIETPUTS FORMATION

> Op die mens en sy sonde
> sy die Here genadig;
> O, Vaalrivier, gly stadig.
>
> -- Herman Charles Bosman
> in "Hyperion"

Record of the Rietputs Formation

River terraces and alluvial fills are present in most landscapes, providing a commonplace record of past geomorphic environments. However interpretation of alluvial chronologies is neither simple nor straightforward. It seems that rivers possess relatively few morphologic responses to the myriad of possible environmental changes (cf. Fairbridge, 1968: 1124-1138; Gregory and Walling, 1973: chap. 7). Stream networks are sediment and fluid delivery systems that either erode or deposit, depending upon complex relationships between fluid supply, sediment supply, and channel characteristics. In addition, these responses can be observed at different scales of space and time. Clearly each suite of terraces and fills requires careful analysis in its own right. The interpretative alternatives in the Lower Vaal Basin include both climatic stimuli and tectonic-lithological responses. Fortunately, in this case, the morphology of the Pre-Karoo Surface and the regional, litho- and pedo-stratigraphy of the alluvial fills limit the possible explanations, so that an economical, interpretive model can be constructed.

The record of the Rietputs Formation comprises morphologic, depositional, and pedogenic features which form an impressive regional mosaic in the modern landscape of the Lower Vaal Basin. This regional patterning is summarized in Table 10-1. It is evident that similar, broadly synchronous events occurred along similar reaches of several streams in the region. This alluvial history and its explanation are the focus of this chapter.

Dating the Rietputs Formation

The duration of particular geomorphic styles and the tempo of the changes between them are important explanatory constraints in evaluating a history of geomorphic environments. Consequently the apparent ages of the ancient alluvia are a first consideration. The Rietputs and early Riverton formations have not yet presented opportunities for radiometric dating. Their ages can only be approximated through external correlation of the included fossils and artifacts. In addition, some of the geological events themselves imply generalized lengths of time. Only the later Riverton Formation is within the effective range of ^{14}C dating.[1]

As argued in Chapter 4 (above), a general age for the Wedburg Terrace of 4.5-2.5 m.y. B.P. is provided by a small collection of battered, mammal fossils (Helgren, 1977). In addition no artifacts

---

[1]The temporal relations of the later members of the Riverton Formation are becoming better understood as research progresses in the Lower Vaal Basin. Deposition of Member V appears to span 4500 - 1300 B.P. The alluvial (later) phase of Member IV apparently spans about 15,000-13,500 B.P., while the colluvial phase is earlier, possibly beginning about 17,500 B.P. (Butzer, in preparation). Member III was deposited somewhat more than 40,000 years ago. These approximate dates are corroborated by correlation with the environmentally sensitive, tufa cycles along the Gaap Escarpment (Butzer, et al., 1978), but datable materials from the alluvial fills themselves remain rare and usually pose contamination problems.

TABLE 10-1

RECORD OF THE RIETPUTS FORMATION

STAGE	VAAL RIVER MAIN CHANNEL	VAAL TRIBUTARY FEATURES
1	Major valley incision and valley widening. Wedburg Terrace abandoned.	Tributary valleys incise. Scarps of Wedburg Terrace begin retreat. Rubefied paleosols develop on some Older Gravel outcrops.
2	Deposition of Rietputs Member A (massive valley-fill of coarse gravels). Channel lengthening at Vaal-Harts confluence.	Tributary channels aggrade coarse sediments derived from erosion of Older Gravels and tillites. Massive interdigitation of Gaap Escarpment tufas with fluvial sediments along Vaal-Harts valley. Shale plains re-grade to the local base-levels provided by the Rietputs-A fill.
3	Valley and valley-fill incision. Widespread erosion of Rietputs-A deposits.	Tributary channels incise. Rubefied paleosol develops on Rietputs-A surfaces. Corrosion and solution of tufa surfaces. During later phases, throughflow calcretes intrude Rietputs A.
4	Aggradation of Rietputs B (mosaics of gravel with sandy interbeds in channels, sandy floodplain margins).	Continued erosion of Rietputs-A fills. Coincident calcretion of Rietputs-B sediments. Massive fluvial sands first appear in Lower Vaal Basin. Cyclic sedimentation at Riverview Estates and Winter's Rush. Floodplain deflation.
5	Erosion of Rietputs-B deposits and creation of associated terrace morphology. Deposition of Rietputs C, lining the modern channel. Channel lengthening at Vaal-Harts confluence.	Erosion of earlier Rietputs alluvia. Throughflow calcretes develop in valley alluvia.
6	Deposition of Member I, Riverton Formation.	Widespread erosion of sandy slopes along river valleys. Dune activity in valley bottoms (?).
7-15	Multiple episodes of floodplain incision and aggradation of Riverton Members II to V.	Non-repetitive sequences of pedogenesis, hillslope erosion floodplain deflation, and incision of tributary channels. Surficial reworking at Older Gravel sites.

HARTS RIVER	RIET RIVER	ORANGE RIVER
Major valley incision and widening. Wedburg Terrace scarps begin retreat.	Major valley incision and widening.	Valley incision and widening. Wedburg Terrace abandoned.
Deposition of Rietputs-A sandy alluvium and lateral tufas from the Gaap Escarpment. Shale plains re-grade to the fill surface.	Deposition of Rietputs-A coarse alluvia (preserved at DeKalk, Riverside, and Schutsekama). Shale plains re-grade to new floodplain elevations.	Rietputs-A equivalent sediments accumulate. Local re-grading of shale plains.
Rietputs-A deposits are incised. Scarps of Rietputs-A terraces begin retreating. Widespread corrosion and solution of tufas.	Rietputs-A deposits and shale bedrocks are incised. Scarps of Rietputs-A terraces begin retreating.	Rietputs-A channelways are incised. Scarps of Rietputs-A terraces begin retreating.
Evidence of Rietputs B preserved by lateral planation below Rietputs-A terrace, forming erosional benches. General calcretion.	Rietputs B (sands and gravels) deposited, with contemporaneous calcretion. Massive calcreted sands at Schutsekama. Subsequent lateral planation, creating Rietputs-B erosional benches.	Rietputs-B deposition. Subsequence lateral planation creating Rietputs-B erosional benches.
Incision of channelway. Corrosion and solution of calcreted Rietputs-B surfaces.	Major channel incision. Deposition of gravel beds (beneath Riverton Formation in Riet valley). Solution and corrosion of Rietputs-B calcretes.	Channel incision. Deposition of low-level gravels on DeHoek and Mazelsfontein Channel lengthening at Orange-Vaal confluence.
Deposition of Member I, Riverton Formation.	Deposition of Member I, Riverton Formation.	Deposition of Member I, Riverton Formation.
Aggradation of Riverton Members II to V.	Aggradation of Riverton Members II to V.	Aggradation of Riverton Members II to V.

have been recovered from primary, Older Gravels deposits.[2] The succeeding erosional phase, which begins the Rietputs record, saw the Vaal valley deepened (up to 50 meters) and widened (up to 8 kilometers, near Douglas) as large volumes of Dwyka sediments, Older Gravels, and assorted hillslope colluvia were removed from pre-Karroo valleyways. An interval in the order of a million or more years seems implied.

A collection of mammal fossils that now numbers over 1000 specimens has been retrieved from the Younger Gravels and Riverton Formation by diamond diggers and other casual collectors (Cooke, 1949; Wells, 1964; Helgren, 1977). Unfortunately more than half of the collection is undescribed, and fossil proveniences are either unknown or uncorrelated to the stratigraphy presented here. Most of the specimens come from Rietputs C and seem to be of later Middle Pleistocene age, although older and younger elements are also present. When the collection is viewed as a whole, it is apparent that several evolutionary stages, as well as several faunas and faunal spans probably are represented. Consequently these fossils presently allow only generalized dating inferences, but hold considerable promise if descriptions and taxonomic revisions are completed.

A large number of artifacts also have been discovered in semi-primary and derived contexts in the Rietputs and early Riverton formations (Helgren, 1978; Malan, 1947; van Riet Lowe, 1952; Mason, 1962; Cole, 1961). Unfortunately little temporal detail is provided since all the artifacts are referable to the Early Stone Age (Acheulian) tool-making tradition. Artifacts are infrequent in Rietputs A, but can be readily recovered from many Rietputs-B and

---

[2]However stone tools are present on the surface at many Older Gravels sites, reflecting repetitive, raw-material quarrying until later Stone Age times.

particularly Rietputs-C outcrops, as well as from gravel lags in Member I and at the base of Member III of the Riverton Formation. Handaxes from contexts in the Riverton Formation are often smaller and associated with a large number of smaller flake and blade tools including some evidence of Levallois technique, suggesting late-Acheulian ("Fauresmith") variants (for a typological discussion see Humphreys, 1969; Sampson, 1974). However the extant artifact collections are dominated by handaxes and cleavers (Cole, 1961), artifacts whose temporal seriation is controversial at best. Occurrences of Early Stone Age artifacts in East Africa have been dated to as old as 1.4 m.y. B.P. by K/Ar controlled magnetostratigraphy (Hay, 1976), so that Rietputs A could be this old. A terminal age for the Acheulian is provided by a uranium-series date of about 200,000 years on a lacustrine (pan) limestone at the Rooidam site, southwest of Kimberley (Butzer, 1974b, Szabo and Butzer, in press). This pan setting stratigraphically may compare best to Member II of the Riverton Formation.

Additional temporal relations are implied by the morpho-, litho-, and pedo-stratigraphies themselves. Rietputs-A deposition locally totaled over 30 m of coarse gravels and coincided with accretion of massive tufas more than 10 km from their sources at the Gaap Escarpment. Such depositional magnitude is without parallel in the later geological history of the region and a long temporal span is implied. At some point during the succeeding erosional phase, a rubefied paleosol developed on Rietputs-A surfaces, and still later a calcrete paleosol developed on the channel margins. Again a lengthy interval seems implied. Rietputs-B deposition is relatively unimpressive in mass compared with Rietputs A, and a shorter interval seems indicated. However the lag deposits of Rietputs C, with its high density of fossils and artifacts, suggests a long interval with only limited geomorphic change.

In summary then, the impression obtains that:

(1) The erosional stage before Rietputs A may represent a period of perhaps 10^6 years around the Pliocene-Pleistocene boundary;

(2) Rietputs-A deposition may span a few intervals of 10^5 years during the early Pleistocene;

(3) The erosional stage between Rietputs A and B may span several intervals of 10^5 years during the early Pleistocene;

(4) Rietputs-B aggradation may span several intervals of 10^4 years during the late Lower Pleistocene or early Middle Pleistocene[3];

(5) Rietputs-C deposition may span several intervals of 10^5 years during the Middle Pleistocene;

(6) Deposition of Member I of the Riverton Formation may span a few intervals of 10^4 years during the late Middle Pleistocene; and

(7) Many intervals characterized by little net geomorphic change presumably occurred between these stages and during interruptions within them.

Consequently the Rietputs Formation apparently records a series of fluctuations in geomorphic environments during the early and mid-Pleistocene. Deposits of comparable age are rare in the interior of southern Africa, and the Younger Gravels seem to comprise a nearly unique record of geomorphic, environmental trends that span the

---

[3]For discussions of Middle Pleistocene dating and stratigraphic problems see Butzer and Isaac (1975).

time between the australopithecine cave-breccias of the Transvaal (e.g. Butzer, 1976; Vrba, 1975) and the better understood environmental changes of the Upper Pleistocene and Holocene (e.g. Butzer, *et al.*, 1978).

Past Interpretation of the Rietputs Formation

The earlier investigators of the Younger Gravels of the Vaal River only considered deposits above Delportshoop, with particular and appropriate focus on Windsorton. In addition, they emphasized the depositional record of the Vaal, neglecting the informative tributary facies and evidence of intervening pedogenesis. Individual explanations were expectably based not only upon the litho- and morpho-stratigraphies perceived but on the conceptual frameworks popular at the time (see also Chaps. 1, 4, and 5 above).

Söhnge and Visser (1937) identified three gravel aggradations (Younger Gravels I, II, and III) which were each preceded by an interval of erosion and succeeded by an interval of sand accretion. To explain this sequence they postulated cyclic fluctuations in precipitation, with the maximum erosion occurring with the precipitation peak, while the gravels and sands were deposited sequentially as precipitation (and presumably streamflow) gradually decreased. Additional important elements in their interpretation were the presumed "barriers" to river downcutting provided by the reaches in andesite topography above Windsorton and below Barkly West. These bedrock thresholds were supposedly lowered sequentially while regional denudation proceeded, with the sporadic Older and Younger Gravels a partial consequence of episodic (and dramatic) changes in local stream gradients.[4] Nonetheless a "rainfall curve" was the focus of their conclusions.

---

[4]This "barrier hypothesis" apparently owes its origin to Stow (1872; 1874). Many aspects of this theory appear nonsensical today.

Cooke (1947) accepted Söhnge and Visser's stratigraphy with several minor modifications, and he at least partly recognized that the "barriers" were features of the Pre-Karoo Surface. He also adopted a climatic, i.e. rainfall interpretation and identified the Younger Gravels interval as the "Younger Pluvial", a period distinctly wetter than the relatively humid climates responsible for aggradation of finer sediments covering the Younger Gravels. His conclusion featured another "rainfall curve."

In 1952 C. van Riet Lowe incorporated Cooke's hypothesis into his final re-evaluation. He performed further minor surgery on the earlier stratigraphy (see Chap. 5 above) and suggested a South African pluvial chronology. Again gravels were presumed to be deposited in very wet phases, while sands and silts were deposited in less humid environments.

Partridge and Brink (1967), looking only at the deposits near Windsorton, explicitly abandoned climatic explanations altogether. They recognized no internal stratigraphy in the Younger Gravels and failed to distinguish between them and the Riverton Formation at Windsorton and Riverview Estates. They further inferred that all the highly variable facies as well as the different paleosols were the expectable response of a river in a semiarid environment. By implication the aggradation of the Younger Gravels and Riverton Formation was seen as the product of infrequent, high magnitude events, thought to be an integral part of the modern hydrologic regimen. By way of explanation for the Older and Younger Gravels they suggested possible combinations of presumed tectonic and stream-capture effects, placing particular emphasis on the upstream

---

However it is clear that valley width characteristics--generally related to features of the Pre-Karoo Surface--are important in determining relative heights of terraces along the different reaches of the lower Vaal River, even though late-Cenozoic breaches in authentic thresholds of the Pre-Karoo Surface remain to be discovered.

passage of cyclic knickpoints in the spirit of L. C. King's Morphology of the Earth (1968). Supposedly generated by dramatic, ancient sea-level changes, these knickpoints were presumed to have migrated up the rivers, possibly splitting at the "barriers", and leaving rock-cut terraces in their wake. On the scale of the Lower Vaal Basin, King's model of landscape development is little more than an interpretive metaphor, that can only be accepted on faith.[5] In any case it is dispensed with in Chapter 9 above.

The principal contributions of this report are a new, region-wide stratigraphy of the Younger Gravels features, a detailed appreciation of the geological context of these deposits, and the interpretative model presented below. It is clear that none of the earlier interpretations has important explanatory value today.

Towards a New Interpretation of the Rietputs Formation

An alluvial fill is the record of a set of superimposed relict floodplains, reflecting an interval of net but not necessarily continuous or homogeneous aggradation along a river valley. The upper surface of a fill ideally preserves the height of the final floodplain before an interval of valley downcutting. Alternatively, the upper surface of such a fill could represent an erosional bevel produced at a much later date. Rock-cut surfaces at the same level as the alluvial fill are the product of coincident hillslope erosion, which persistently removed rock waste and soil products toward the channel. In this way the processes of rock weathering, soil formation, and hillslope erosion provide the dominant sources (and initial sedimentary attributes) for the clastic debris found in the alluvial fill. The unconformities between alluvial fills, or along the

---

[5]There seems little room in King's model to accommodate alluvial deposits, even as relatively transitory features.

bedrock at the base of an alluvial fill, record erosional phases when the main stream and its local tributaries incised earlier alluvia and bedrock surfaces, removing part of that alluvial record and leaving behind limited and usually transitory fill on eroded surfaces. Tributary channels grading to these new channel levels erode, so potentially destroying earlier terraces, even though soil formation on valley surfaces may proceed apace. The Rietputs and Riverton formations apparently record many such aggradational and erosional phases (see Table 10-1). However, explanations are far from simple.

For net alluviation to occur, the geomorphic environment must persistently provide more debris to the channelway than can be removed downvalley by the processes of sediment transport operating in the channel. Consequently a change to net alluviation can be the result of increased sediment supply or of reduced effectiveness in sediment transport or both. For valley incision to occur, there must persistently be less debris provided along the channelway than channel transport processes can regularly remove. The resultant excess energy is then directed toward erosion, favoring channel widening and/or deepening. A decrease in sediment supply or an increase in the effectiveness of the sediment transport, or both, is implied. In this sense alluvial fills and intervening erosional phases are the highest-order categories in the history of a river valley, averaging across innumerable cut-and-fill events that occur constantly as channels adjust to continuous (small-scale) variations in fluid and sediment supply.

Unfortunately, the interpretation of complex alluvial fills is difficult, because both the processes of sediment supply (i.e. rock weathering, soil development, hillslope erosion) and those of sediment erosion and transport in alluvial channels are interrelated, not only to each other (since the results of a first process usually effect the character of later processes) but also to other factors in the wider geomorphic environment. Furthermore, since progress in geomorphic systems is step-wise, there are time lags both in the delivery of hillslope debris to the valleyway and in the movement of the alluvial debris downstream. This emphasizes how the processes responsible for a cumulative history of floodplain aggradation and incision have variable magnitude, frequency, spatial location, and temporal context. More explicitly, in regard to the geomorphic work recorded by an erosional or incisional phase, such changes: (1) did not occur with constant intensity through time; (2) may have been accomplished by an assortment of events with variable magnitudes, durations and frequencies; and (3) were not uniformly distributed across river valley landscapes at any particular time. It is further reasonable to expect that many intervals in an alluvial history are characterized by equilibria in valley morphology and sedimentology,[6] that such episodes of equilibria will be more frequent during the later parts of both aggradational and erosional phases,[7] and that the transitions between phases of net aggradation and net erosion represent formidible thresholds which can only be

---

[6]To some extent these intervals of equilibria should be preserved in the pedo-stratigraphic record.

[7]This impression is supported by the existence of rock-cut terraces, that imply most of the responsible lateral erosion occurred when stream levels were persistently near the same level. Additional support is provided by simple aspects of the valley geometry. The upper parts of valley cross-sections are wider than the lower parts. Consequently a change in floodplain level near the end of an aggradational phase would require proportionately greater mass-transfer than at an intermediary stage, and further aggradation becomes more costly to the fluvial system. Similarly, when valleys are strongly incised, the slopes adjacent to the channels are longer, and tend to be steeper, implying progressive increases in sediment supply as an erosional phase proceeds; thus continued valley erosion becomes more difficult.

crossed by major (high-order) changes in the geomorphic environment. In short, the establishment of detailed, causal relationships to explain alluvial histories is rarely possible with the current state of geomorphologic research, so that identification of general themes and trends in the formative geomorphic environments must continue to provide an acceptable interim goal of interpretation.

Three general alternatives are available for interpretation of the Rietputs Formation. The first considers the Rietputs Formation as the product of temporally and spatially random events, primarily of high magnitude and short duration. The second alternative explains these deposits as the product of several structural-lithological effects. And the third sees the Rietputs Formation as the product of changing geomorphic environments induced by regional climatic changes. The conclusion of the present study is that the climatic alternative provides the only viable interpretative hypothesis.

The role of temporally and spatially random events of high magnitude (catastrophies) as an explanation in Natural History has a long tradition that needs no amplification here. However, if causal, systematic relationships describe modern geomorphic environments (as suggested by the last two decades of geomorphic research), the same must have been true in the past. The complex of features preserved in the Rietputs Formation provide ample evidence not only of geomorphic gradualism,[8] but an empirical record of cumulative, non-comparable sequences that are often unalike. For example, the Rietputs-A deposits at Winter's Rush

---

[8]Gradualism itself seems to be a relative idea. In particular, empirical evidence that alluvial landforms do not respond significantly to rare, high-magnitude floods is presented by Dury (1974), but this notion seems best applied to humid rivers flowing on relict, fine-grained fills. The evidence from small, arid drainage basins is decidedly different (Baker, 1977). Both examples are probably "gradualist".

include an interbedded floodplain vertisol, implying progressive aggradation of this pedosedimentary unit. Also diagnostic of "slow" aggradation are the laminated tufas interdigitated with Rietputs-A alluvium at the Ulco pump, as well as the finely bedded, Rietputs-B sands and Riverton sediments at Riverview II and Windsorton I, each of which preserve penecontemporary root-casts and root hollows. The many paleosols and gravelly lag horizons, in addition to the archeological and paleontological antiquity of the deposits, provide equally impressive evidence that progressive geomorphic changes were the rule, even though trends were often reversed over longer timespans. The random event hypothesis for the Rietputs Formation must be considered inappropriate.

Three structural-lithologic hypotheses can be offered for the Rietputs Formation, namely that the related features reflect:

(1) changing areal dimensions of the Vaal River basin;

(2) tectonic deformation (tilting) and variable relief in the drainage basin; and/or

(3) knickpoint retreat or destruction.

It will be argued here that none of these hypotheses are viable.[9]

The Vaal River basin has decreased in area since the mid-Cenozoic. However, no sudden or important changes can be ascribed to the period of the Rietputs or Riverton formations. On the eastern margin of the basin, in the eastern Transvaal and Orange Free State, the steep-gradient streams draining down the east-facing Drakensberg have continued to extend their headwaters at the expense of the Vaal system throughout the later Cenozoic, possibly aided by tectonic tilt to the east. This has resulted in complex captures by Natal streams of the lowest-order drainageways along this divide. The modern divide is a 10 to 60 kilometer-wide plateau,

---

[9]Some of these alternatives have been discussed in Chapters Three and Four above.

poorly integrated into either basin, and which probably contributes relatively little stream discharge to either (see Wellington, 1943). Estimating the flow loss from this rim of the Vaal basin during the time of Rietputs and Riverton deposition is extremely difficult, if not impossible. In any case, such captures involved no more than minor channels, so that the proportional loss in discharge to the lower end of the basin would have been negligible.

Similar capture events are also recorded in the geomorphic history of the northern margin of the Vaal basin, now marked by the Witwatersrand and the parallel "Transvaal-Griqualand Axis" of du Toit (1933). The complex captures that have occurred along this apparently warped margin of the basin are indicated by the courses of the more ancient Older Gravels (pre-Wedburg Terrace) of the western Transvaal. The pattern of these deposits demonstrates the loss of drainage by the Vaal to the Limpopo and Molopo systems (see Williams, 1932; Mayer, 1973). However, these captures must significantly pre-date aggradation of the Rietputs Formation.[10] All the available evidence indicates that the Vaal basin today is substantially similar in size and shape to the Vaal basin during the times of Rietputs sedimentation.

Evidence for tectonic deformation of the Vaal River basin is unimpressive. As noted in Chapter 3 above, tectonic deformation in the Lower Vaal Basin since the Dwyka glaciation has been geomorphologically insignificant. Post-Dwyka tectonic adjustments within the lower basin are mainly recorded by 1 to 4 centimeter-wide

---

[10]Where the "London Run" of Older Gravels (between Schweizer-Reneke and Bloemhof) joins the modern Vaal valley, these deposits stand on valley shoulders well above the Rietputs-equivalent terraces and consist only of colluvial lags; the primary alluvial bodies have been eroded away. In addition, no primary artifacts have been associated with these deposits (van Riet Lowe, 1952).

joints cutting across striations on roches moutonnées of Ventersdorp andesite. Displacements up to 20 cm are known. These joints may relate to late-Dwyka isostatic rebound, the emplacement of late-Karoo dolerites, the emplacement of post-Karoo kimberlites, or/and isostatic adjustments initiated by the removal of a few thousand meters of later-Karoo sediments by Cenozoic erosion. Some joints could equally well be part of pre-Dwyka joints systems that were plastically "erased" by the Dwyka glaciers, only to re-establish themselves at later dates. In any case such jointing is unimportant to the argument here and the lower basin must be viewed as internally cohesive and stable within the recent geological past. Whatever tectonic forces may have affected the lower basin operated universally.

The possibility must be considered that deformations of a subcontinental scale tilted the upper basin during the Quaternary, increasing relief and so affecting river behavior. A steeper regional gradient would presumably favor initial valley incision.[11] Such movement has been postulated along Vaal-Limpopo-Molopo watershed (du Toit, 1933; Mayer, 1973), but the dates of the presumed flexures are unknown[12] and evidence within the Lower Vaal Basin is incompatible with such a proposed N-S tilting. If tilting were to account for the geomorphic cycles established here, then first of all, the chronology of multiple episodes of large-scale aggradation and downcutting would require that the whole drainage basin be tilted

---

[11]The potential impact that a tilt of a fraction of a degree would have on floodplain dynamics a few hundred kilometers downstream remains debatable.

[12]It is worth noting that the evidence for such tectonism (gentle warps in presumably once-flat surfaces, and stream captures in low-order drainage networks) could be explained economically by differential erosion of variably resistant rocks.

repeatedly, a supposition that severely strains the demonstrable empirical framework. Secondly, Rietputs deposits and terraces can be systematically traced not only along the southwest-trending reaches of the Vaal, but also on the northwest-trending reach from Gong-Gong to Delportshoop as well as along the westward-flowing Riet River and the northwestward flowing Orange.[13] Thirdly and most importantly, the shapes in the Pre-Karoo Surface (see Chap. 3) seem to record a regional slope already established in Paleozoic times and which has changed little since. These facts are wholly incompatible with the deformation hypothesis.

Knickpoint destruction or retreat also does not explain the record of the Rietputs Formation. Presumably channel breaching of a threshold in resistent rock should cause rapid incision immediately upstream, until the knickpoint is smoothened and eliminated.[14] The modern channels in the Lower Vaal Basin are situated within Paleozoic valleys of the Pre-Karoo Surface, or shaped in argillites above the Pre-Karoo Surface. In either case, any knickpoints would have developed in non-resistant Dwyka rocks and would therefore have been quite ephemeral. Intrusions of Karoo dolerite could have provided more durable knickpoints, but the field evidence is unimpressive. At several sites along the Riet and Harts rivers there are dolerite dykes, but they appear to have been breached with only

---

[13]The gravel deposits along the lower Vaal River have often been offered as proof of deformation along the Witwatersrand. This argument is based on the fallacious assumption that the acceleration of hydraulic processes by regional tilting causes channel scour and the production of coarse clasts. Instead, gravels are provided to channels by hillslope weathering and erosional processes in particular, local bedrock settings and specific, regional climatic (or hydrologic) opportunities that need have nothing to do with tectonism.

[14]Valid empirical evidence for knickpoint retreat over long distances has yet to be presented (see Chap. 9 above).

minimal depositional and morphologic responses. Large sills seem to have affected valley width more than channel elevation. In any event, for such an explanation to be viable, more than a dozen, now-missing knickpoints would be essential along the several streams. Furthermore, each of these knickpoints would require breaching throughout the system at precisely the right time to produce any one episode recorded in the Rietputs or Riverton formations, let alone a temporal scenario involving multiple regional events that would generate the entire mid-Pleistocene to Holocene alluvial history of the basin. Finally, the alluvial fill structures themselves can not be explained by a simple mechanism of sequential knickpoint change.

Having eliminated possible random and structural-tectonic explanations for the Rietputs Formation, the alternative of climatic change remains, if only by default. While any sequence in Natural History can be explained by climatic changes, the terraces and deposits along the channels in the Lower Vaal Basin as well as inferences that can be made from modern analogs actually provide a robust explanation.

An Environmental Scenario for Alluvial Fills

Phases of floodplain aggradation as well as phases of floodplain incision can occur in both humid and arid climatic settings (cf. Schumm, 1968; Baker and Penteado-Orellana, 1977). Evidently floodplain behavior is not predicated simply on the nature of subcontinental climatic zones. Instead floodplain dynamics are intimately linked to sediment supply and textures, and to magnitude-frequency aspects of the formative processes (Wolman and Miller, 1960; Baker, 1977), which produce typical depositional features, channel morphologies, and paleosols. These processes are linked to climatic systems in anything but a one-to-one way. Perhaps the only unifying theme is the frequency of formative discharges and the character of fluid delivery from the basin slopes. Humid environments appear to provide formative discharges relatively frequently (every few years or less), with precipitation reaching the channels largely via throughflow and ground water discharges. Arid environments provide such discharges only after longer intervals (after several years or decades), with precipitation primarily reaching channels via overland flow. Given this situation, it seems best to focus on the geomorphic environments that condition or predicate stream behavior, rather than on climatic regions. A generalized model of geomorphic environments governing many subtropical stream systems is proposed here (Table 10-2). The basic characteristics can be enumerated under the headings "Humid Incision", "Arid Aggradation", "Arid Incision", and "Humid Aggradation". Also important are the "Transitions" from one of these geomorphic environments to the next.

Humid Incision is a consequence of frequent, high discharges which remove sediment from valleyways more rapidly than debris is supplied by hillslope erosion and tributary streams. Humid-incision streams have low-sinuosity channel plans,[15] closely adjusted to bedrock structure. Channel cross-sections are relatively narrow and deep, with high hydraulic radii.[16] Floodplains are narrow (i.e. largely occupied by the channel), and bedrock is often exposed along the channel. Deposits on the floodplain (which may have any texture) expectably are ephemeral, but fragmentary segments may be buried by a later aggradational phase. Slopes beyond the floodplain have

---

[15]Sinuosity is defined as the ratio of stream length to valley length (Schumm, 1963).

[16]Hydraulic radius is the cross-sectional area of a channel divided by the distance along the wetted perimeter (Morisawa, 1968:29).

well developed soils. Modern streams following this model frequently occur in well-vegetated, subtropical landscapes where high-magnitude rainfalls are frequent, and precipitation reaches the streams mainly via throughflow and groundwater discharges (Garner, 1974, Chap. 5). In addition such streams are present in nearly all high relief landscapes, even those with subhumid climates.

Arid Aggradation is typical where infrequent, high-intensity precipitation events over open vegetation mosaics generate erosive overland flows, moving more debris to the valleyways than the even less frequent, formative stream discharges can transport downstream. As a result the streams aggrade. Necessary are readily erodible slopes, provided either by non-resistant bedrocks, or by thick soils remaining on the interfluves from an earlier humid environment. If the abundance of fine-grained sediments on the slopes is related to a non-resistent bedrock, then this depositional style can continue as long as relief is available. When relict soils are being eroded, the aggradational phase will last only as long as fine-textured, soil products are available. Shallow, zonal soils of arid aspect may develop on low-relief interfluves, but true soils will be rare elsewhere. Channel characteristics are probably generalized. High discharges may occupy the whole valley, but low discharges typically are carried in wide and shallow channels with low hydraulic radii.[17] Sediment textures may be critical in determining channel morphologies (Schumm, 1960). Channel plans have moderate sinuosity, and braiding may be frequent at lower flow stages. Deposits are dominated by the coarser fractions of former slope sediments (often sands), since the finer fractions may be lost in suspended load transport

---

[17] Local, tributary channels may incise their upper reaches while aggrading their lower reaches (see Garner, 1974, Chap. 6).

TABLE 10-2

ALLUVIAL FILL ENVIRONMENTS AND TRANSITIONS

From \ To	Humid Incision	Arid Aggradation
Humid Incision	Frequent, high-magnitude events lead to active channel incision. Channels have low sinuosity and high hydraulic radii. Humid, zonal soils develop on the interfluves.	Valley aggrades with sediment eroded from nearby slopes. Channel sinuosity increases and hydraulic radii decrease. Floodplain deflation may be widespread. Humid zonal soils on interfluves are eroded and arid zonal soils develop. Basal unconformities often may be steeply dipping and irregular.
Arid Aggradation	Valleys incise, perhaps destroying earlier alluvial fills. Channels become less sinuous and hydraulic radii increase. Response may be complex until humid, zonal soils develop on the interfluves.	Infrequent, high-magnitude events carry slope debris to fill river valleys. Sediments have "arid" characteristics and variable facies development. Channels have moderate sinuosity and, probably, low hydraulic radii. Shallow, arid zonal soils develop on interfluves.
Arid Incision	Incision continues without creation of definitive new features. Humid, zonal soils develop on the interfluves	Unlikely to occur without an intervening humid phase when soils develop, or a change in the sediment-delivery system caused by availability of a new, less-resistant rock-type. Channel sinuosity probably increases, hydraulic radii may decrease. Soil development trends are unchanged.
Humid Aggradation	Valleys incise, perhaps destroying earlier alluvial fills. Channel sinuosity decreases. Soil development trends change little.	Remodeling of floodplains by attributes related to arid aggradation. Channel sinuosity and hydraulic radii may decrease perhaps after initial incision. Soil development on interfluves shifts from humid to arid types. Intervening unconformities may be gradual and subtle.

Arid Incision	Humid Aggradation
Unlikely to occur except when humid and arid climates alternate rapidly, and deposition will typically intervene. Perhaps no definitive features preserved. Soil development on interfluves reverses from humid to arid trends.	Valleys aggrade and humid-geomorphic floodplains develop. Channels lengthen, becoming more sinuous. Basal unconformities may be steeply dipping and irregular.
Valley fill incision, perhaps of limited magnitude in response to decreasing sediment supply. Channel sinuosity decreases while hydraulic radii may increase. Limited soil development on interfluves continues.	Channels and sediments acquire humid-geomorphic attributes. Channels probably become more sinuous and hydraulic radii probably increase (as sediments become finer?). Soil development on the interfluves shifts from arid to humid trends. Intervening unconformities are smooth and perhaps gradual.
Infrequent, high-magnitude events incise relict alluvia. Adjacent hillslopes are nearly sediment free. Channels have low sinuosity and variable hydraulic radii. Arid, zonal soils may develop on low-angle interfluves and relict fills.	Valleys aggrade and humid-geomorphic floodplains develop. Channels lengthen, becoming more sinuous. Soil development on the interfluves shifts from arid to humid trends. Basal unconformities may be steeply dipping and irregular.
Humid-geomorphic floodplains are incised. Channel sinuosity decreases, and hydraulic radii may increase. Soil erosion is active on the interfluves. Arid soils may develop on slopes with low rates of erosion.	Frequent, moderate-magnitude events aggrade floodplains; channel morphologies are related to sediment type. Vertisols develop on floodplains, while zonal soils develop on interfluves.

or deflation during channel desiccation. Depositional residues are equally generalized with only limited facies diversity. Horizontal bedding is predominent with only limited preservation of the cross-stratification developed during peak discharges. The ephemeral "rivers" of the southern Kalahari provide generalized, modern analogs for this model.

Arid Incision is the consequence of infrequent, high-intensity or high-magnitude events in landscapes relatively resistant to erosion and yielding only limited hillslope sediments. Arid incision is the expectable response to continuing erosion of relict soils that had previously supported "arid aggradation". In particular, as the supply of relict soil sediments decreases, the frequent, high-intensity precipitation continues, now only slowly eroding the "new", more resistant slopes, and incising the relict alluvial debris along the valleyway.[18] The incised channels have low to medial hydraulic radii and low-sinuosity plans adjusted to structure. Arid, zonal soils may develop on the surfaces of relict, alluvial fills as well as on low-angle surfaces along the interfluves, perhaps providing the only objective contrast to an interval of "humid incision". Many desert channelways in high-relief terrains or in basins with particularly resistant bedrocks provide examples of this model.

Humid Aggradation is characteristic in a variety of intermediate climatic settings, with a considerable range of synoptic attributes. The basic theme is a relative surplus in sediment delivery to the valley floor where it is sorted and shaped by high-frequency channel

---

[18]Such an incisional episode probably has limited potential for bedrock erosion unless the formative environment is very persistent.

processes as the perennial channels adapt least-work (most efficient) morphologies. Such floodplains have been the focus of fluvial geomorphic research for the past two decades. Dury (1976) suggests that formative stream discharges reoccur at less than 2 year intervals. Most frequently low-intensity soil erosion provides silts and clays, but coarser sediments may be provided in frost environments or by weathering of unusual bedrocks. The resulting channel characteristics tend to reflect sediment type (Schumm, 1960). Where bedrock does not constrain adjustments in channel plan, channel lengthening in relation to the channels of any previous phase is typical. In the case of clays or silts, the channel network may be highly sinuous and even meandering in plan. Large changes in channel location often occur during events of formative magnitude or greater, and fragments of relict channels may be present. "Silt/clay" channels are relatively narrow and deep with high hydraulic radii. Depositional residues are dominated by horizontally stratified, fine sediments, while relict channel deposits may exhibit rapid changes in texture and stratification preserved in discontinuous beds (e.g., Beerbower, 1964). Floodplain vertisols may be preserved while zonal soils develop on the interfluves. Several shallow and muddy, meandering tributaries of the uppermost Vaal basin provide modern analogs. Aggradational sand and gravel streams, where crude detritus is abundant, are relatively wide and shallow, with low sinuosity channels that often braid; resulting deposits preserve an assortment of channel and flood basin facies. Many modern streams in sandy landscapes and on high altitude/latitude piedmonts respond to analogous discharge and sediment-supply environments.

<u>Transitions</u> from one incisional or aggradational phase to another are induced by major changes in the geomorphic environment. Transitions often preserve diagnostic sediments, landforms, unconformities,

and/or paleosols along the valleyways and interfluves. A transition from arid aggradation to humid aggradation is one such a case. Other transitions may leave no diagnostic features, or no record at all may be preserved. A transition from arid incision to humid incision provides such a case. The general phase and transition attributes are summarized in Table 10-2. Most alluvial-fill networks only imperfectly preserve a sequence of several transitions, so that several alternative interpretations are possible (see also Cooke and Reeves, 1976, pt. 1). Also, it must be emphasized again that the transitions of this model represent only the highest-order thresholds in alluvial histories (for a discussion of alluvial thresholds, see Schumm, 1973; 1977), since alluvial landscapes can accommodate important, smaller-scale changes in discharge and sediment supply through rapid but transitory adjustments in channel characteristics.[19]

For any phase in an alluvial-fill chronology, this geomorphic-environmental model provides at least one explanation. For example, the modern floodplain incision and gullying typical in most semiarid landscapes (Butzer, 1971) can be referred directly to a transition from "humid aggradation" to "arid incision" caused by historic groundcover destruction, related to intensified agricultural settlement (overgrazing and plowing). The opened vegetation mat inhibited soil infiltration, increasing the relative proportion of overland flow, leading to greater but less sustained peak discharges, with gullying of footslopes and general channel incision, even though

---

[19]Adjustments in channel cross-section, size, and plan in response to altered sediment supply and regimen characteristics have been studied on broad alluvial surfaces by Schumm (1968) and by Baker and Penteado-Orellana (1977). Their research points to the possible diversity of such morphologic changes, which are the expectable product of the poorly understood dynamics of sediment supply. The possible adjustments in narrow valleys lined by resistant bedrock are more problematical, and await investigation.

annual rainfall statistics may have changed very little. Such a syndrome may be reinforced by increased intensity of precipitation events (Cooke and Reeves, 1976), perhaps also in part a consequence of vegetation change. The opened vegetation may favor more intense heating of land surfaces, more frequent and energetic convection cells, so providing more opportunities for thunderstorms within suitable macro-circulations. While this particular hypothesis can neither be proven nor refuted, the models of "humid aggradation" and "arid incision" provide a useful explanatory framework that fits the available data.

Enthusiasm for the models presented above must be tempered since their interpretative power can be exaggerated. A first problem is that the alluvial environments are identified in terms of sediment supply and stream discharge relationships, which are only indirectly linked with the patterns of synoptic climate. Geomorphic environments are at least as sensitive to bedrock structure and weathering preferences as evapotranspiration and runoff ratios. A second difficulty is that, despite decades of intensive study devoted to alluvial landscapes, the requisite empirical data to test this model are not yet available. This situation extends beyond the obvious problems of short, temporal perspectives. In particular the association of specific alluvial landscapes with the debris or unconformity geologically recorded in an alluvial-fill complex remains imprecise. Furthermore, quantitative research has been directed to analysis of streams flowing across wide alluvial surfaces, while streams in narrow valleys and bedrock channels with little sediment have been neglected. In addition, the phase trends and transitions of modern streams are difficult to identify by modern testing methods, which emphasize repeated short-term adjustments in discharge and sediment supply.[20]

---

[20]Perhaps the ultimate test for this model would be descriptions of phase transitions "caught in the act". Schumm and Lichty (1963)

A third problem is that local and regional effects in sedimentary behavior become blurred for trans-environmental rivers like the Vaal, which flow from humid to arid regions.[21] It also must be emphasized that the alluvial environments refer primarily to events along the river valley, whereas geomorphic processes and trends on the interfluves (e.g., pans on the shale plains or tufas along the Gaap Escarpment) may reveal environmental transitions of different magnitudes and tempos.

Despite these reservations, the alluvial environmental models presented here can be applied to the lower Vaal River, even though the sedimentary data are fragmentary, few relict channels are preserved, and the rivers of the Lower Vaal Basin are often confined to narrow valleys of the Pre-Karoo Surface.

Environments of Rietputs A

The valley incision that followed the Wedburg Terrace, the aggradation of Rietputs A, as well as the valley incision following Rietputs A (Stages 1, 2, and 3 of Table 10-1) can be readily identified as phases of "humid incision", "humid aggradation", and "humid incision" respectively. In addition, the calcreted, tributary channels eroded in Rietputs A suggest that a semiarid environment was effective before deposition of Rietputs B.

---

apparently describe a sequence of transitions (or a complex of perturbations which ultimately will lead to one phase transition) in the recent history of the Cimarron River of Kansas. Yet the segments of these changes that will persist in the fill record remain a matter of speculation.

[21]For example, the historical Nile River in Egypt was a "humid-aggrading" stream in terms of the model, despite the aridity of the adjacent Saharan landscapes. Sediment ponding and flow regulations at the Aswan High Dam have begun to turn it into a "humid-incising" stream.

The magnitude of the valley incision through Karoo bedrocks preceding deposition of Rietputs A is impressive, and explanation in terms of frequent, high-magnitude discharges in a humid geomorphic environment seems necessary. Incision along even minor tributaries indicates that high discharges were arriving both from the upper basin and from the interfluves of the lower basin. Perhaps the rubefied paleosols found at many Older Gravels outcrops (see Chap. 4) should be correlated with this episode of valley incision. Although the only sediments preserved from this incisional phase are the "Basal Fines" at Riverview I, the relevant paleo-Vaal was transporting the largest clasts provided from erosion of tillites, Older Gravels, and dolerites, so indicating competence and capacity well beyond that of the modern Vaal. Perhaps this phase of bedrock incision represents the most humid environment in the late-Cenozoic history of the Lower Vaal Basin.

The morphologies of valley incision have been simulated in a scale model by Shepherd and Schumm (1974). Their flume experiments, using high velocities and steep slopes, produced incised channels that were relatively straight and locally "overdeepened". The pre-Rietputs valley of the Vaal, particularly between Warrenton and Winter's Rush, closely mirrors this situation, with relatively straight valleyways related to structural preferences in the Pre-Karoo Surface.[22] In addition, the unusually deep "potholes" at Windsorton and Gong-Gong are similar to the "overdeepening" observed in the flume. However, primary erosion of these andesite valleyways relates to discharges during Paleozoic times.

---

[22]To initiate incision Shepherd and Schumm (1974) tilted their flume, but downcutting undoubtedly could also have been induced by adjusting other variables. Since the model of Shepherd and Schumm simulated "bedrock" with a mixture of sand with 19% "silt-clay", a variety of non-scaled attributes were probably introduced. Incision of resistant rock seems different by at least an order of magnitude from incision of an alluvial fill.

The deposition of Rietputs A is a geomorphic theme without parallel in the subsequent geologic history of the region. Never has the Vaal deposited such large masses of gravels nor have the Gaap Escarpment tufas reached the Vaal channel in such volumes.[23] A significant interlude of "humid aggradation" seems indicated, even though the synoptic climates may have been less humid, providing more erratic discharges than the preceding episode of "humid incision". Rietputs-A alluvia are best preserved at Windsorton, from Gong-Gong to Winter's Rush and near Delportshoop (Chap. 5 above). Rietputs-A deposits downstream of Delportshoop and along the Riet and Harts rivers are shallow and discontinuous, but they do provide generalized, collaborating evidence.

The typical Vaal alluvia of Rietputs A are crudely-stratified, pebble to boulder-grade gravels, set in a compact matrix of inhomogeneous sand. Thin, discontinuous lenses of finer sediments are rare. Unfortunately the alluvial dynamics of such large, poorly-sorted clasts are probably the least understood of any fluvial sediments. Streams transporting comparable sediments in modern landscapes typically braid, with the coarse, bedload debris shaped into lateral (transverse) and longitudinal bars (Schumm, 1960; Fahnestock, 1963; Williams and Rust, 1969; 1972; Church, 1972).[24] Such channels are wide and relatively shallow. Floods usually submerge the whole valley bottom, while at lower flows the bars rise above the water surface to form both mid-channel islands and sub-channels in assorted sizes. The modern studies (particularly

---

[23]Along the Gaap Escarpment the incisional phase preceding Rietputs A is represented by an important interval of erosion (Butzer, et al., 1978). The absence of tufa deposition can be referred to high discharges along the escarpment that were undersaturated in carbonates.

[24]These studies of streams reworking earlier glacial sediments provide an intriguing historical parallel for the Vaal River.

Williams and Rust, 1969) also describe a variety of fine-grained, ripple and dune-form structures in low-flow channelways. If this analogue for Rietputs A is appropriate, these fine sediments tend not to persist. Perhaps bed-form diversity was restricted by the narrow valley confines of the Lower Vaal Basin. Amid shale and tillite bedrocks, fine-grained sediments should have been an important part of the load. The Rietputs-A depositional style of large, crudely stratified clasts is probably the product of deposition during receding peak floods, with the matrix sands trapped as the coarser materials come to rest or subsequently filling the open interstices. No doubt such deposits are resistant to later fluvial erosion. If this description is appropriate, then the Rietputs-A deposits at Windsorton and Winter's Rush record only the terminal stages of many formative floods.

Rietputs-A alluvia along the Vaal below Delportshoop and in the Riet and Harts valleys are too fragmentary to permit modelling of channel morphologies. The sands and gravels at Selene on the Vaal and at DeKalk on the Riet suggest wide, shallow channels with gravel splays pulsing downstream. This is much like the modern Orange below the Vaal confluence. The other exposures are little more than lag deposits and speculation is inappropriate.[25]

That the environment of Rietputs-A deposition was more humid than present is verified by the tributary limestone tufas. These tufas have their origin along the Gaap Escarpment where ephemeral springs and streams discharge carbonate-saturated waters along the scarp face. Enduring tufas are not being deposited now (Chap. 2 above), and a significantly more humid climate would be required to generate tufa carapaces along the escarpment today (Butzer, 1974a;

---

[25]Rietputs deposits are only fragmentarily preserved between Windsorton and Barkly West where the low gradient suggests that non-durable, non-rudaceous deposits may have once been prevalent.

Butzer, et al., 1978). Tufa flowstones reached the Vaal-Harts channel during Rietputs-A aggradation. Other informative tributary facies of Rietputs-A age are the colluvial gravels which merge with valley alluvia in the steep, andesite terrains. These suggest that Rietputs-A deposition was linked to increased sediment supply resulting from greatly increased hillslope erosion. The most economical interpretation for this increased erosion is frost-induced, mass-wasting processes. Such processes appear to be recorded by lenses of frost-shattered limestone in the more massive tufa deposits along the Gaap Escarpment (Butzer, et al., 1978). Consequently, Rietputs "humidity" probably was a response to both high precipitation and decreased evapotranspiration. In any case, the coarse gravels of the Vaal document repeated and frequent high-magnitude discharges from the upper basin, while both the tufas and the phenomenal competence of surficial sediment mobilization along the valley sides demonstrate a moist climate and persistent, high discharges in the lower basin.

The channel incision following deposition of Rietputs A was accompanied by the development of a rubefied paleosol on the adjacent interfluves. A lengthy interval of "humid incision" with an environment at least as humid as present is indicated. Presumably the patterns of precipitation and stream discharge that typified Rietputs-A deposition persisted even after the rapid hillslope erosion of Rietputs-A times had ceased. Channelways eroded within the Rietputs-A fills eventually were calcreted towards the end of this incisional phase, signaling a progressive transition to a semiarid or even arid environment prior to deposition of Rietputs B.

Rietputs-A Paleo-discharge

Paleo-hydrologic and paleo-hydraulic inferences can be constructed both from relict, channel morphologies (e.g., Schumm, 1972;

Dury, 1976), and from sedimentary characteristics (e.g., Jopling, 1966; Baker, 1974; Church and Gilbert, 1975; Clague, 1975). In addition, channel morphology and sedimentary data sometimes can be combined to produce more comprehensive estimates (e.g., Cotter, 1971). Even though the results of these studies are tantalizing, caution is necessary. The morphologic inferences are developed from statistical comparisions among modern streams and, unfortunately, accurate analogy from modern streams to alluvial-fill structures remains difficult. In addition, the necessary preconditions for reconstruction often cannot be fully met or even determined from the incomplete record preserved in alluvial fills. Similarly, the estimates from sedimentary data are based on engineering procedures, which often do not cope well with the diversity of natural systems. With these cautions in mind, an attempt can be made to reconstruct the hydrology of Rietputs-A deposition.

A reasonable estimate is possible for Rietputs-A deposition near Winter's Rush. Several hydraulic factors are already known or can be easily estimated for this reach of an ancient Vaal River.

Energy gradient (S) can be approximated from the average slope of the Rietputs-A terrace in the Gong-Gong to Winter's Rush reach, where $S = 0.00164$.

Particle size (D) is directly measurable. The average, median diameter of the 10 largest clasts from the Rietputs-A sample at Winter's Rush indicates $D = 0.099$ m.

Depth (d) can be approximated when both S and D are known. Baker (1974: 104) argues from a dimensionless shear standpoint that:

$$d = 0.099\ D/S \text{ (metric units)} \quad \text{(equation 1)}$$

when the specific weights of quartz and relatively clear water are assumed.[26] Consequently the predicted depth during formative discharge is d = 5.98 m.

Channel roughness (n) or the resistance to flow provided by the channel itself can also be estimated. Roughness provided by clasts on the bed (Chow, 1959: 205-206) is approximated by:

$$n = 0.038\ D^{.166} \qquad \text{(equation 2)}$$

and for D = 0.099, n = 0.025. However, this approximation neglects several factors, including probable bank vegetation and the lateral limits on channel width provided by the andesite walls of the pre-Karoo valleyway near Winter's Rush. Consequently, an n = 0.03 is assumed here.

With d, S, and n available, the average, channel-forming velocity (Va) can be determined from the Manning Equation (Church and Gilbert, 1975):

$$Va = \frac{R^{.66}\ S^{.5}}{n} \qquad \text{(equation 3)}$$

where R is the hydraulic radius, defined as the ratio of the channel cross-sectional area to the wetted perimeter. For very wide channels, R is approximated by d, and equation (3) becomes:

$$Va = \frac{d^{.66}\ S^{.5}}{n} \qquad \text{(equation 4)}$$

and for Rietputs A at Winter's Rush,

$Va = 4.45$ m/sec.[27]

---

[26]Building on the arguments of Church (1972), that the sediment packing patterns, sediment mix, and particularly temperature-related viscosity must be included, Clague (1975: 231) proposed: d = 0.184 D/S, for Wisconsinan discharges in the Rocky Mountain Trench. This prediction is considered inappropriate here since Rietputs-A gravels preserve virtually no imbrication and, at 29° latitude, temperature effects need not be considered.

Channel width (w) is extremely difficult to approximate from exposures in alluvial fills. Such fills reflect net deposition both of migrating, principal channelways and of ephemeral tributaries, while channel structures are rarely preserved (for further discussion see Jopling, 1966; Schumm, 1972). No relict channel structures are known from Rietputs A. However, an estimate of 360 m is provided by the width of a narrow, bedrock-bounded reach immediately above Winter's Rush, where Rietputs A is not preserved. This 360 m is about 50% of the valley width at Winter's Rush and Waldeck's Plant, and about 15% of valley width at Windsorton. This estimate is compatible with the examples of Schumm (1968), and reasonable for the width of the "live bed" channel during Rietputs-A floods, which probably inundated the whole valleyway near Winter's Rush.

With estimates of Va, d, and w available, the formative, peak discharges (Qp) during periods of flow capable of moving the largest particles can be approximated by:

$$Qp = Va\ d\ w \qquad \text{(equation 5)}$$

so that Qp for Winter's Rush is about 9700 m^3/sec.

Finally, mean annual discharge (Qm) can also be estimated. Schumm (1972) collected and compared data on a large number of rivers flowing in relict alluvia, and found that for wide, shallow streams in relatively humid landscapes where formative discharges were occurring about every two years (as is likely for the Rietputs-A Vaal)

$$Qm = \frac{w^{2.43}}{18\ \left(\frac{w}{d}\right)^{1.13}} \qquad \text{(in English units--c.f.s.).}$$

---

[27]This predicted average velocity is compatible with the bed velocity of about 1.75 m/sec necessary for mobilization of clasts with intermediate diameters near 10 cm (Helly, 1969; Hjulström, 1935; Mavis and Laushey, 1949; Fahnestock, 1963).

For the Winter's Rush case, Qm can be computed to about

15,800 c.f.s. or 450 m^3/sec.

On this basis the average annual discharge for the Rietputs-A Vaal above the Harts confluence was about:

1.42×10^{10} m^3/yr.

No great accuracy is claimed for this number, but the general order of magnitude is probably valid. The average annual discharge of the modern Vaal at Riverton (1909-1948) is 3.095×10^9 m^3/year (see Table 2.5). Simple comparison suggests that Rietputs-A Vaal had a discharge four to five times that of today.[28] The limestone tufas that merge with Rietputs A along the lower Harts River and below the Vaal-Harts confluence suggest that even greater discharges were characteristic in these reaches.[29] Similar discharge patterns probably characterized the incisional phase following deposition of Rietputs A.

Environments of Rietputs B

Informative exposures of Rietputs B are located at Riverview Estates (Windsorton) and at Winter's Rush (see Chap. 5). Elsewhere in the region, deposits are shallow and fragmentary; typically,

---

[28]These reconstructed discharges are minima only. Dury (1976) has pointed out that Schumm's "form ratio" estimators significantly underpredict discharges for the largest rivers in humid environments. In addition, if the depth estimator of Clague (1975) and an n of 0.025 are used, predicted annual discharge more than doubles. The channel-width estimate also is a minimum, since the measuring site may have been a narrow reach constricted by the pre-Karoo bedrocks.

[29]In principle, paleo-hydrological estimates should be possible for the lower Harts River where the preconditions for Dury's model (1976) appear to be more fully met. Unfortunately, reasonable estimates for Rietputs-A channel width and channel depth could not be obtained.

only a thin gravel lag rests on a calcreted terrace surface. The depositional theme at Windsorton and Winter's Rush is episodic gravel mobilization, but with stream competences more frequently in the sand grades. Gravel clasts from Rietputs-A terraces were being reworked in shallow, sandy channels. The cyclic layers of finer debris at both sites record desiccation along tributary channels and in peripheral floodbasins. The Vaal discharge to the Orange may have been very low. The image of greater aridity is reinforced by the persistent calcretion of the valley alluvia. No tufa correlates are known at the Gaap Escarpment, and no significant tributary alluvia were discovered. The impression obtains that the Rietputs-B environment was arid, with highly periodic discharges, provided primarily from the upper basin. A phase of "arid aggradation" is indicated. Apparently both the upper and lower Vaal basins, as well as the Riet-Modder drainage of the High Veld were more arid than at present.

Paleo-hydrological speculations are inappropriate for Rietputs B. Clearly the largest discharges were transporting coarse gravels, and these floods may have been as great as the formative discharges of Rietputs A. However, the flood frequency was probably much lower, and the available, paleo-hydrological models cannot be applied to the infrequent, discharge patterns typical of arid landscapes.

Origin of the Hutton Sands

Even though the Rietputs-B deposits are unimpressive, they signal a major change in the geomorphic history of the Lower Vaal Basin. First appearing during this phase are the Hutton Sands, which created permeable land surfaces that drastically changed the subsequent patterns of hillslope erosion. Panchromatic air photos and LANDSAT images clearly show that Hutton Sand veneers are distributed primarily to the south and east of the modern Vaal channel (van Rooyen and Burger, 1974a; 1974b), indicating genesis through floodplain deflation by northwesterly winds. The modern, locally discontinuous, distribution of Hutton Sands, and the orientations of the now-relict dune ridges seem to imply multiple episodes of secondary remobilization. The primary source for the Hutton Sands is indicated by the red, sandy bank sediments of Rietputs-B age preserved at Winter's Rush. Circumstantial evidence for this association is also provided by the hillslope erosion of Rietputs-A times, which would have been impossible in a landscape covered by thick, permeable sands. Floodplain deflation must be added to the roster of arid characteristics for the Rietputs-B environment.

Red, sandy detritus is readily produced in several geomorphic settings. For example, modern footslope colluvia in the andesite uplands of the Lower Vaal Basin are dominated by red sands, as are slope debris around oxidized outcrops of Dwyka sediments (as at Wedburg, see Chap. 4). In addition several Older Gravel deposits (i.e. Holpan, Proksch Koppie, Brandt's Kleim, see Chap. 4) record the limited availability of red sands during the later Cenozoic. However the volume of the Hutton Sands exceeds the apparent capacity of any historic or modern source within the region, and a source outside the lower basin must be sought.[30]

Looking outside the lower basin there are two apparent sources for the Hutton Sands, and both can be supported. Red surficial

---

[30]Piaget (1963) carried out heavy mineral study of many relict, red sands in the northern Cape Province and northeastern Orange Free State. In the southern Kalahari he was able to associate some sand bodies with immediate bedrock sources, but the heavy mineral assemblages of the sands to the southeast of the Vaal did not yield systematic proveniences. Part of this interpretative problem relates to the erratic sand lithologies in the Dwyka tillites. However, mixing in fluvial and eolian environments could also explain this heavy mineral pattern.

sediments with an important proportion of sand have been present in the upper basin since at least Wedburg Terrace-times, as verified by the red colluvia at the Transvaal australopithecine sites (Butzer, 1976) and at the Virginia fossil locality (Butzer, 1973). The first explanation for the Hutton Sands sees them derived from oxisols (or oxic ultisols and alfisols), which developed in the southern Transvaal and northern Orange Free State during the apparently more humid climates of Rietput-A times.[31] With the onset of Rietputs-B aridity, soil erosion was probably widespread, moving slope debris down the Vaal basin. With alluvial sorting, as the fines were carried away in suspension, the sandy channels would have been subject to deflation during periods of low flow.

The second explanation sees the Hutton Sands ultimately coming from the Kalahari basin, to the north of the Molopo River. This explanation requires multiple episodes of eolian transport of sands into the western Transvaal, where they were eventually tapped by the northern tributaries of the Vaal (see also Flint and Bond, 1968). This explanation is supported by the relict sand veneers of the western Transvaal and by the Hutton Sands along the Harts River valley north of Spitskop.[32] Both alternatives may have been significant in providing the sandy bed-loads of the Rietputs-B Vaal.[33]

---

[31]Presumably large volumes of red sands could have been generated quickly through weathering of Ecca and Beaufort sandstones in the upper basin.

[32]The Hutton Sands along the upper, western margin of the Vaal-Harts interfluve west of Holpan can be explained through weathering and erosion of the andesite bedrocks. However, Hutton Sands are unusual on the shale plains of the lower Harts valley. It is tempting to speculate that the Harts valley above Spitskop (in the modern Vaal-Harts irrigation project where Hutton Sands are abundant) served as a sedimentary basin for Hutton Sands during Rietputs-B times with only infrequent, low-sediment flows reaching the Vaal.

Environments of Rietputs C

By the end of Rietputs-C times the bedrock morphology of most of the modern river valleys in the Lower Vaal Basin had been established. Rietputs C records the transition to an incisional phase, following aggradation of Rietputs B. Deposits of Rietputs-C age along the Vaal preserve only channel facies and are found cemented within or near the modern channel from Windsorton to beyond Schmidtsdrif. No equivalent tributary facies are known, and Gaap Escarpment tufas did not reach the floodplain. Deposits below the Riet-Vaal confluence are not verified. The artifacts and mammal fossils recovered from Rietputs-C beds (Helgren, 1977) suggest relatively brief mobilization and prompt burial, during infrequent episodes of channel scour. Presumably the paleontological materials initially accumulated along the channelway during low-flow periods, in small ephemeral, flood basins and on gravel bars, prior to limited transport and burial during floods.

Several geomorphic environments probably were formative in the valley incision and aggradation of Rietputs C. The initial incision following Rietputs B is best referred to an interval of arid incision, either as the sources of the Rietputs-B alluvia were

---

[33] A problem related to the appearance of the Hutton Sands is the near termination of exorheic drainage in the "Sand Veld" of the northwestern Orange Free State (see Wellington, 1945). The modern terrains in this area are dominated by chains of pans, which only partly connect to the Vet-Sand and Modder river basins through groundwater discharges. Some such drainage networks (vloers) seem to owe their origin to ponding upon arrival of the Hutton Sands. Discontinuous, subtle valleyways (some with vloers) lead out of the northwest Orange Free State toward the Vaal at both Windsorton and above Barkly West. Perhaps stream discharges from the modern Sand Veld contributed to the Vaal of Rietputs-A times and earlier.

terminated or as discharges of fluid and sediment were both reduced.[34] Continuing semiarid conditions in the lower basin are implied by calcretion of Rietputs A and B deposits and probably also by the episodic, intermediate-competence scour-and-fill probably responsible for preservation of the fossil mammals. However, persistent discharges in some sort of humid environment during this interval are also suggested by (a) the mobilization iron and manganese as represented by the oxide cements common to the many outcrops of Rietputs C, (b) the apparent channel lengthening that occurred at the Vaal-Harts confluence,[35] and (c) the surficial corrosion of calcreted Rietputs-B surfaces. Consequently Rietputs C seems to imply a long interval of intermediate, subarid to subhumid environments along the Vaal, with average conditions similar to those of the semiarid present.

Along the Harts River, Rietputs C is recorded by a modest incisional phase, and geomorphic environments like those of the Vaal basin are most reasonable. However, along the Riet River, complex, coarse-grade alluvia are preserved beneath the Riverton Formation. In addition, along both the Riet and Orange Rivers, the Rietputs-C incisional phase is much more impressive than along the

---

[34]The paleo-climatological implications of the alternative geomorphic environments inducing incision are decidedly unclear in this case. If incision was induced only by a decrease in sediment supply from the upper basin, an interval of increasing humidity in the upper basin with thickened vegetation mats and deepening soils is implied. Alternatively the landscapes in the upper basin may have "run out of" readily erodible surficial sediments, and continuing aridity is implied. If incision was induced by decreased discharges of fluid and sediment, then progressive, increasing aridity in the upper basin is indicated.

[35]Rietputs-C deposits are nearly 800 m wide at the Vaal-Harts confluence. This is more than 10 times the modern channel width.

Vaal, and compares best with the incision episode that followed the Wedburg aggradation along the Vaal. These differential magnitudes of incision may simply be related to relative ease in opening Paleozoic valleys filled with shales, opposed to valleys filled with coarse gravels. However, climatological responses may also have been operative. Perhaps the climatic mosaic over southern Africa during Rietputs-C times was characterized by significantly more humid conditions over the Riet and Orange basins than the Vaal basin. By implication this impressive interval of incision along the Orange and Riet valleys may record a long interval of humid incision, in terms of the model outlined above.

Environments of the Riverton Formation

The Riverton Formation has been noted here in cursory fashion only, and full identification of the later mid-Pleistocene to Holocene environments that these deposits record must await later publication. However some general themes can be outlined.

The massive sands of Members I and III seem to indicate intervals of arid aggradation with massive sediment inputs from the upper basin, and perhaps signal renewed episodes of deflation from sandy floodplains, as well as additional cycles of deposition and remobilization of Hutton Sands. The finer sediments, sinuous relict channelways, and vertisols of Members II, IV, and V suggest variable intervals of "humid aggradation",[36] while the inter-member incisional phases all have both "arid" and "humid" aspects. No doubt the Riverton Formation preserves many environments both more humid and more arid than present. The full dimensions of later Quaternary environment changes in the Lower Vaal Basin will

---

[36]A minor tufa reached the Vaal channel near the Ulco pump during deposition of Member IV alluvium.

be revealed when data from the valley alluvia, Gaap Escarpment tufas, and ephemeral lake (pan) terraces are fully integrated.

APPENDICES

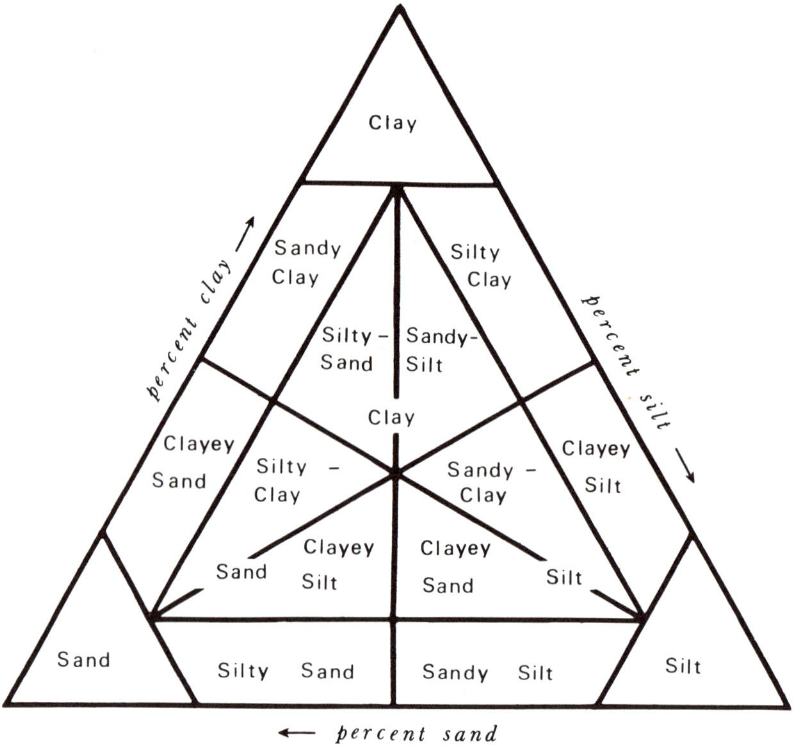

Fig. A-1. Texture classification (after Link, 1966).

APPENDIX A

LABORATORY PROCEDURES AND SEDIMENT DATA

Sediment Analyses

Laboratory analysis of sediments is an essential to regional geomorphic study. It allows more accurate lithological description and facies interpretation, and provides numerical data that can be compared from one exposure or landscape segment to the next. There are logistical problems, however. In dealing with a large area, each segment of which has complex sedimentary units, only a limited number of samples can be processed, and the data so generated provides no more than a selection of numerical parameters that are not fully representative of the full range of lateral and vertical variation within a sedimentary unit. For this reason the sedimentary data presented here are exploratory rather than exhaustive, seeking to provide broad, semi-quantitative descriptions, rather than refined statistical analyses. Accordingly, the terminologies employed tend to be informal, and refined sedimentological parameters have deliberately not been presented. The data presented in Tables A-1 and B-1 explicitly provide a comprehensive but generalized, descriptive framework.

Each sediment sample was selected subjectively as the most representative of the particular stratum or soil horizon. Eventually more than 300 samples were analyzed for textural, chemical and lithological attributes at the Paleoecology Laboratory at the University of Chicago, the Physical Geography Laboratory at the

University of Toronto, and the Geomorphology Laboratory at the University of California, Davis. The modal sample size was 80 g, and portions of the original sample as well as sand residues have been retained. Characteristics of the samples reported in the text are summarized in Table A-1.

The size classes (of long axes) employed here follow a modified Wentworth scale:

cobbles	60 to 250 mm
pebbles	4 to 60 mm
granules	2 to 4 mm
coarse sand	0.56 to 2 mm
medium sand	0.21 to 0.56 mm
fine sand	0.064 to 0.21 mm
coarse silt	0.037 to 0.064 mm
medium silt	0.006 to 0.037 mm
fine silt	0.002 to 0.006 mm
clay	under 0.002 mm

In a general way materials coarser than 2 mm are referred to as gravel. Finer textural categories used in the subsequent chapters follow the simple descriptive system of Link (1966; Fig. A-1). Clay minerals are distinguished from "clay" as a textural term. Silt and sand have specific grade connotations, while silts and sands (plurals) refer only in a general way to sediments of this general calibration.

Sorting is here used to describe the concentration of the materials under 2 mm within the size classes listed above:

very well sorted	- 55% or more (by weight) in the largest size class
well sorted	- 45-55% in the largest size class
moderately sorted	- 40-45% in the largest size class
poorly sorted	- 35-45% in the largest size class
very poorly sorted	- 30-35% in the largest size class
unsorted	- less than 30% in the largest size class

The primary mode is the dominant size-class, while the secondary mode is the second largest size-class.

Structure, consolidation and cementation of soils and sediments follow the specific definitions in Soil Taxonomy (Soil Survey Staff, 1975, Appendix 1), as do the soil horizon descriptions. Colors are given on samples in the dry state according to the Munsell Soil Color Charts (1954). Methods of gravel shape analysis used here are described in Appendix B. Rounding of sand and gravel clasts is the percent of smoothed convex surface, where 0-10% is designated as angular, 11-20% as subangular, 21-40% as subrounded, 41-60% as rounded, and greater than 60% as well rounded.

Soil and sediment horizon boundaries follow a system modified after the Soil Taxonomy (pp. 642-463), which recognizes both distinctness and boundary topography. Distinctness is abrupt if less than 25 mm wide; clear is 25-60 mm wide; gradual is 60-125 mm wide; and diffuse if more than 125 mm wide. The surface topography is smooth, if nearly a plane; wavy if undulations are wider than their depth; irregular if undulations are deeper than they are wide; and broken if the boundary is discontinuous.

The laboratory procedures included the following:

(1) Acid Bath. Nearly all the samples were bathed in a 20% HCl solution, and subsequently washed and dried. The excepted samples were those with a high proportion of coarse, calcrete detritus. Acid treatment was necessary to remove the frequent carbonate cements in the clay to sand fractions, and provided the "HCl solubles" (by weight) statistics in Table A-1. The "HCl solubles" are dominated by calcium carbonates (always over 95%, and usually over 98% of the total on cross-checked samples), but the procedure also resulted in the loss of organic colloids, colloidal forms of silica, iron oxides (originally sand grain coatings), and magnesium carbonates. Magnesium carbonate (dolomitic) cements were present only rarely, and did not seem to vary systematically either

with apparent age or depositional environment. Carbonate contents of most samples also were determined in a Chittick gasometric apparatus, which only proved meaningful for samples with low carbonate contents. The "HCl colubles" listed in Table A-1 for those samples less than 10% are an average of the "bulk" and "Chittick" results.

(2) Texture. Texture was determined using the standard hydrometer analysis, supplemented by sieving of the fraction coarser than 37 microns. After the acid bath the dried samples (usually 50 to 70 g) were peptized in a 5% solution of sodium pyrophosphate, and at least 12 hours later stirred for 5 minutes in a commercial blender prior to the hydrometer measurements. After hydrometer analysis the residue was sieved through 2000, 500, 210, 63, and 37 micron sieves. Selected samples were also sieved at 1000 and 125 microns. The sieved fractions were dried and weighed, and sand grain lithologies, morphologies (roundness and shape, see Powers, 1953), and surface textures (frosting, polish, Fe-coating) were semi-quantitatively determined.

(3) pH. pH was obtained electrometrically with a Beckman pH meter, using a paste of a 10-20 g sample and distilled water. Equilibrium intervals between preparation and determination were from 10 to 20 minutes.

(4) Amorphous Silica Determination. Cementation by subcrystalline (amorphous) silica (SiO_2) and possible silicates (e.g. allophane) proved to be a particularly difficult problem on some samples, particularly the "calcretes" in Dwyka shale terrains. No systematic, temporal or spatial patterning to this cementation was discovered; however, Dwyka tillite matrices subjected to the procedures described above consistently yield high proportions of this "amorphous silica". Biogenic SiO_2 may have been an additional source, as well as chemical weathering of the local volcanic rocks. Optical resolution on thin sections did not prove possible

either (see Figs. 4-7, 4-10, and 4-11). Silica geochemistry at one atmosphere or less and temperatures from 0° to 40°C remains largely uncharted (Davis, 1964; Krauskopf, 1959; Siever, 1957).

In the acid bath samples yielded variable proportions of gelatinous floccules of such "amorphous silica". Upon washing and drying these floccules hardened and became part of the mechanical-analysis fraction. After wet sieving and drying most of these amorphous-silica bits (which included small porportions of clay- and silt-sized debris) became part of the > 37 micron fraction. The proportions of these bits in each size class were then semi-quantitatively determined under the binocular microscope. For sieve-fractions with more than 10% amorphous silica bits, the > 37 micron was bathed in a 10% solution of NaOH, which dissolved most of the amorphous silica, or at least reduced such aggregates to sub-37 micron sizes. The > 37 micron fraction was again wet sieved, dried and weighed. After re-examination under the microscope, the original amorphous silica proportion (the amount lost in NaOH solution plus the percentage remaining) of the > 37 micron fraction was determined. This amount was assumed to be proportional to the < 37 micron fraction and the statistic for total amorphous silica percentage of the sample (including the HCl soluble fraction) was computed. No great accuracy is claimed for this method, but the results are significant in a relative sense. For amorphous silica proportions of 5-10%, the accuracy is only ± 2%, while for proportions above 10% the accuracy is probably ± 5%. Equally important is that such secondary aggregates do not distort the distribution curve of the detrital particles.

TABLE A-1

SEDIMENT DATA

Sample No.	Munsell Color	Texture[a]								pH	HCl Solubles (%)	Amorphous Silica (%)
001	2.5 YR 3.5/7	0.6	2.2	19.5	55.9	4.0	7.6	3.2	7.0	7.8	1.5	0
002	2.5 YR 3.5/7	---	8.8	18.0	49.6	6.0	8.8	3.3	5.5	---	2.1	0
010	5 YR 7/2.5	0.1	0.7	1.4	1.5	5.7	52.1	20.7	17.8	7.8	2.2	7
011	10 YR 7/6	0.3	3.0	11.4	41.4	8.6	24.5	6.1	4.6	7.2	2.5	1
012	3.7 Y 8/2	0	0.1	0.1	6.0	28.5	37.4	13.3	14.6	8.2	1.3	1.5
013	10 YR 7/3.5	0	0.1	0.1	1.8	16.1	56.1	14.0	11.8	8.5	2.6	0.5
014	5 Y 7.5/2	0	0.1	0.3	48.6	12.5	16.1	9.6	12.8	8.0	1.3	1
015	10 YR 7.5/3.5	4.4	4.9	8.6	34.2	10.2	26.8	4.6	6.3	7.8	4.5	5
016	5 Y 7/1	0	0.1	0.1	29.6	22.8	26.3	11.3	9.8	8.1	1.3	Trace
017	10 YR 7/3	6.1	4.8	4.5	38.7	11.0	24.4	5.8	4.7	7.7	2.6	15
018	10 YR 6/8	---	14.7	45.7	24.8	3.3	7.8	1.4	2.3	---	0.8	0
019	2.5 Y 7/2	---	20.8	35.4	17.6	3.9	9.7	5.9	6.7	8.5	1.1	0.5
020	10 YR 7/2-4	---	13.5	43.8	20.0	4.0	6.5	6.5	5.7	8.9	29	Trace
021	10 YR 7/2.5	---	37.1	41.3	9.3	2.5	3.8	3.0	3.0	8.5	1.6	Trace
022	10 YR 7/3	---	9.2	25.3	30.5	6.6	13.1	5.5	9.8	7.8	45	3
023	10 YR 6/3	---	19.8	37.9	21.8	4.0	9.0	3.7	3.8	8.5	3.2	1
024	10 YR 7.5/1	---	16.5	25.2	25.8	5.4	13.0	6.5	7.6	7.9	48	1.5
025	10 YR 6-7/4	---	28.0	40.6	15.5	2.6	9.1	0.8	3.4	8.5	1.6	4
026	5 YR 6/8	3.6	8.9	16.9	47.1	6.3	10.5	3.7	3.0	8.5	5.7	11

027	10 YR 8/6	1.7	14.8	44.6	18.4	4.4	6.8	4.0	5.3	8.4	1.2	1
028	7.5 YR 7/4	---	47.5	19.6	18.2	3.5	6.4	1.6	3.2	8.1	2.0	0
029	10 YR 7/4	---	19.8	33.8	16.8	4.2	13.4	5.0	7.0	8.2	0.8	6
030A	10 YR 8/2-4	---	1.9	12.3	56.6	10.4	13.0	2.7	3.1	8.8	33	19
030B	10 YR 8/2-4	---	7.3	14.6	29.3	6.2	31.8	2.0	8.8	8.8	33	42
031A	10 YR 8/2-3	0.1	1.9	11.6	35.9	9.0	22.3	7.8	11.4	8.7	65	14
031B	5 YR 4.5/5	1.3	4.1	16.0	38.4	7.8	21.6	5.4	5.4	8.7	10.3	10
032	5 YR 3/5	0.4	5.6	16.3	43.5	6.1	16.2	6.3	5.6	8.1	2.1	1
033	5 YR 4/4	0.1	0.8	10.2	40.6	13.5	20.9	11.6	2.3	8.0	7.9	1
034	5 YR 5/5	0.1	1.9	8.9	37.7	14.3	28.1	5.4	3.6	8.3	29.1	0
035	10 YR 10/1	0	0.8	6.6	34.9	7.3	22.2	10.6	17.6	8.7	75	6
036	5 YR 8/3	0	0.6	4.0	39.1	8.6	33.4	8.4	5.9	8.3	60	13
037	5 YR 10/1	0.1	1.2	7.8	42.0	8.8	19.0	9.1	12.0	8.4	65	12
038	5 Y 7/1	0.8	1.7	10.2	50.8	8.9	16.1	4.6	6.9	8.3	27	53
039	10 YR 4/3	0.1	2.6	11.1	41.7	10.3	23.7	6.3	4.2	8.2	9.3	21
040	7.5 YR 5/5	0	1.9	11.4	48.3	13.1	16.0	5.0	4.3	8.4	22	26
041	7.5 YR 8/7	0	0.5	7.3	52.3	11.0	17.2	4.9	6.8	8.5	4.8	15
042	10 YR 7/3-4	0	0.1	1.2	58.6	14.8	9.9	4.2	11.2	8.2	62	2
043	2.5 Y 7/4	0	0.1	0.7	70.7	12.5	10.1	3.4	2.5	8.4	2.5	5
044	2.5 Y 3/1	0	0.8	6.1	51.5	11.6	15.8	6.3	7.9	8.5	52	3
045	2.5 Y 7.5/4	0.1	0.5	4.8	71.9	6.5	7.6	2.6	6.0	8.3	2.0	7
046	10 YR 7/4	0	1.1	14.2	37.8	6.5	27.4	4.3	8.7	8.6	45	14
047	2.5 Y 8/5	0	0.7	33.0	50.1	4.6	7.1	2.7	1.8	8.3	0.5	0.5
048	7.5 YR gray	0	0.1	19.4	49.8	9.4	10.6	3.2	7.5	8.5	48	0.5
049	8.5 YR 7/4	0.2	6.0	57.3	23.9	3.2	5.3	1.8	2.3	8.3	0.4	0
050	7.5 YR 5/1	0.3	7.4	33.8	40.5	7.2	4.9	1.0	4.9	9.1	42	0.5
051	10 YR 7/3	2.4	21.2	48.9	17.6	2.9	4.1	2.1	0.8	8.6	0.9	4

Table A-1--Continued

Sample No.	Munsell Color	Texture[a]								pH	HCl Solubles (%)	Amorphous Silica (%)
052	10 YR 5/1	0.5	6.3	29.0	42.0	5.9	9.2	2.0	5.1	8.7	44	3
053	10 YR 7/3	0	0.5	3.7	26.1	8.0	37.3	10.7	13.7	8.0	40	16
054	10 YR 7/3	0.6	5.0	44.5	32.5	3.5	8.1	2.3	3.5	---	2.5	11
055	10 YR 7/4	---	18.6	38.8	14.8	5.9	13.2	3.2	5.5	8.2	35	Trace
056	10 YR 8/2	---	13.5	24.4	25.2	4.0	11.7	7.7	13.5	8.2	36	3
057	5 Y 8/3	---	27.6	26.7	21.9	4.6	9.0	3.7	6.5	7.9	1.3	3
058	10 YR 7/3-4	---	33.8	32.2	17.3	3.7	5.3	1.2	6.5	8.0	0.8	Trace
059	10 YR 8/4	---	18.4	40.0	21.0	4.7	10.8	2.7	2.4	7.8	1.1	0
060	5 YR 4/8	---	18.1	15.6	34.8	3.4	16.2	3.4	8.5	8.0	2.8	7
061	5 YR 7/2-3	---	30.1	22.6	16.3	6.5	13.2	7.5	3.8	8.0	57	5
062	10 YR 7/3	---	44.4	15.6	22.1	7.7	7.0	1.6	1.6	7.8	2.4	11
063a	5 YR 5-6/8	6.6	10.1	10.7	48.3	4.4	0.1	0.1	24.7	8.1	2.2	15
063b	2.5 Y 7/2	0.6	13.0	13.2	38.4	11.5	12.2	3.8	7.3	8.2	52	2
064	10 YR 8/3	---	13.8	15.6	38.8	11.4	13.2	4.1	3.1	8.6	53	3
066a	10 YR 8/3	0.6	4.3	8.7	46.1	15.3	15.2	4.5	5.3	7.9	33	15
066b	2.5 YR 6/6	6.2	7.4	8.4	40.5	10.8	8.0	2.7	16.0	8.7	50	1
067	10 YR 8/3	---	42.4	18.2	20.4	6.5	7.1	3.6	1.8	8.1	2.9	5
068	10 YR 8/3-4	---	14.7	12.3	44.4	9.7	6.3	6.3	6.3	8.4	68	3
069	10 YR 7/3	---	22.4	22.4	34.3	5.4	7.7	2.6	5.2	8.0	5.5	18
070	10 R 8/3	0.1	4.1	14.8	51.2	6.8	3.3	3.3	16.4	8.2	76	7
071	2.5 YR 4/8	2.1	7.9	15.2	56.3	6.4	5.5	1.1	5.5	7.1	1.5	0.5

073	2.5 YR 3/8	0.1	0.3	5.3	39.7	16.2	23.8	5.3	9.3	7.8	14.2	11
074	2.5 Y 8/6	0.5	0.2	4.8	10.9	5.6	43.9	10.9	23.2	7.6	16.7	26
075	10 YR 7/3	---	49.2	33.7	9.1	0.2	3.4	1.1	3.3	8.0	1.5	11
076	3.8 YR 5/8	---	13.0	13.2	28.8	10.0	17.1	7.2	10.7	8.4	8.4	14
077	2.5 YR 4-5/6	0	0.2	5.6	48.1	10.3	18.3	7.6	9.9	7.0	2.7	Trace
079	5 YR 6/8	0	0.1	15.4	60.8	10.5	9.0	1.5	2.7	7.9	9.0	15
080	10 YR 6-7/4	0	0.1	10.4	54.6	8.1	13.5	5.7	7.6	8.2	1.0	Trace
081	5 YR 5-6/8	0.1	0.3	10.1	53.9	11.0	15.6	3.3	5.9	8.3	2.3	Trace
082	10 YR 7/4	0.5	0.6	2.0	31.6	15.5	32.6	9.0	8.2	7.8	2.8	3
083	10 YR 7/3	0	0.1	0.5	14.6	14.8	45.5	12.2	12.3	7.8	32	15
084	10 YR 7/4	0	0.1	0.1	9.8	24.6	46.8	9.3	9.3	8.2	2.0	19
085	10 YR 8/3	0	0.1	0.1	58.0	12.8	20.3	3.9	4.8	8.1	1.8	2
086	10 YR 7/4	---	29.4	25.6	27.2	3.2	5.2	3.4	6.0	8.1	1.5	1
087	5 Y 8/4	---	33.4	21.6	14.0	4.4	10.5	5.6	10.5	7.5	4.3	4
088	6 YR 6-7/6	4.8	13.0	24.2	39.1	4.5	4.8	4.0	5.6	8.0	3.5	0
089	5 Y 8/2	0.7	17.2	52.4	19.8	1.4	2.6	1.3	4.6	8.2	1.0	1
090	10 YR 9/2	1.1	15.8	67.0	7.5	1.2	0	3.7	3.7	8.5	44	Trace
091	2.5 Y 8/2	5.0	35.2	45.3	7.0	1.0	1.6	1.6	3.3	8.2	2.0	4
092	2.5 Y 3/4	6.6	42.9	33.4	7.0	1.4	4.4	0.8	3.5	8.1	17	5
093	7.5 YR 6/3	4.2	11.0	29.3	37.5	5.3	5.3	4.2	3.2	8.4	6.1	5
094	2.5 Y 9/2	---	12.2	43.0	26.8	6.8	7.0	3.2	1.0	8.2	7.0	10
095	5 Y 9/2	4.0	58.4	19.6	10.6	0.3	1.2	1.2	4.7	8.6	37	2
096	10 YR 9/2	1.2	43.8	36.0	8.2	2.3	5.4	0.8	2.3	8.8	10	9
097	2.5 Y 8/3	0	0.1	0.3	54.6	6.6	17.1	12.8	8.5	7.9	4.2	10
098	2.5 Y 8/2	0	0	0.1	26.9	17.3	37.3	9.2	9.2	8.0	1.8	4
099	10 YR 7/4	5.1	6.7	15.0	39.1	13.8	15.7	2.3	2.3	7.8	3.8	19
100	10 YR 7/3	0	0.1	0.8	5.8	6.0	54.2	14.6	18.5	7.6	5.1	21

Table A-1--Continued

Sample No.	Munsell Color	Texture[a]								pH	HCl Solubles (%)	Amorphous Silica (%)
101	7.5 YR 8/4	0	1.7	14.2	40.4	4.7	21.0	7.7	10.3	7.7	1.5	12
103	10 YR 7-8/3	---	31.7	20.2	21.3	5.6	11.8	3.2	6.2	7.5	2.3	Trace
104	7.5 YR 6/5	0	0.1	1.1	24.3	12.7	30.9	14.0	16.9	7.7	3.0	11
105a	5 YR 6/10	0	0.1	0.1	85.5	4.7	2.7	1.4	5.5	8.2	2.0	4
105b	7.5 YR 6/3	0.1	1.6	3.0	45.7	13.3	15.1	9.0	12.2	8.4	4.0	Trace
106	10 YR 5/4	0.1	0.1	0.5	11.3	15.0	43.4	14.8	14.8	7.9	5.1	18
108	7.5 YR 6/7	0	0.1	0.3	60.6	13.2	14.8	5.0	6.0	7.8	2.4	2
109	7.5 YR 6/5	0.1	0.1	0.4	9.7	18.6	39.6	14.9	16.6	8.0	7.2	19
110	7.5 YR 6/5	0.1	0.1	0.6	15.1	11.7	47.3	12.9	12.2	7.9	10.3	6
111	7.5 YR 7/7	0	0.1	0.1	18.4	15.8	42.0	13.3	10.3	7.8	2.0	5
112	7.5 YR 6/5	0	0.3	1.4	44.4	16.5	23.6	8.1	5.7	7.8	2.0	6
114	10 YR 6/5	0.1	0.1	1.3	39.4	22.0	23.8	5.3	8.0	7.5	1.8	1.5
116	7.5 YR 7/7	0.1	0.9	21.1	57.2	9.9	8.6	0.7	1.5	8.0	1.5	0
117	10 YR 8/6	0.1	0.6	13.3	49.2	10.5	19.3	3.5	3.5	8.3	2.6	22
118	7.5 YR 6/6	0.1	0.1	10.8	66.0	9.0	10.8	1.3	1.9	8.0	1.4	0
128	N 10	0	0.2	2.7	59.0	11.8	9.2	4.0	13.1	8.0	87	4
129	N 10	0	0.1	0.6	37.6	5.4	21.4	14.0	20.9	8.1	87	4
130	7.5 YR 10/1	0	0.1	0.8	79.4	7.1	2.8	5.6	4.2	---	78	5
138	5 Y 7-8/2	0.6	3.5	32.3	21.7	5.2	17.3	6.7	12.7	8.3	72	5
139	2.5 YR 5/9	0.1	1.0	21.2	53.8	6.8	8.5	3.4	5.2	7.7	1.0	0
140	2.5 YR 4/8	---	21.6	21.9	31.4	3.3	5.4	8.2	8.2	7.5	2.0	Trace
141	2.5 YR 4/8	---	18.4	29.0	32.5	3.0	2.9	7.1	7.1	7.5	2.6	Trace

146	2.5 YR 3/8	---	2.4	14.9	54.3	7.7	11.0	4.5	5.2	7.7	2.1	0
147	5 YR 8/6	---	10.7	24.9	30.2	6.6	15.2	5.8	6.5	8.0	45	14
148	2.5 YR 8/3	0.1	0.7	62.1	19.7	3.1	9.2	1.7	3.4	8.0	4.2	4
149	2.5 YR 10/2	0.1	0.7	50.8	32.8	2.8	7.5	1.9	3.4	---	5.6	4
150	10 YR 8/3	---	17.1	40.8	21.5	4.0	10.7	2.6	3.3	7.9	2.0	4
151	10 YR 8/3	0	1.3	33.6	53.3	2.0	4.5	0.9	4.4	8.0	3.6	5
152	5 YR 5/7	3.2	3.2	16.8	54.4	7.1	5.3	2.5	7.5	8.1	31	Trace
153	2.5 Y 8/1	---	4.8	38.2	28.4	7.6	13.6	3.4	4.0	---	58	5
154	2.5 Y 7/4	---	44.6	26.3	11.7	3.8	7.4	2.2	4.0	7.7	3.4	11
155	5 YR 5/8	---	14.6	33.7	34.0	3.0	8.4	2.1	4.2	7.8	1.3	3
185	10 YR 7/3	0	0.1	3.4	49.1	8.8	20.6	6.3	11.7	---	16	34
186	10 YR 7/4	---	14.6	28.0	25.7	8.8	14.3	2.1	6.5	---	12	27
187	10 YR 10/1	---	19.8	24.3	19.3	9.1	11.7	4.1	11.7	---	59	13
188	7.5 YR 9/1	15.3	6.3	13.2	34.1	0.6	13.8	5.6	11.1	---	72	10
190	10 YR 9/2	0.1	0.1	7.4	37.0	15.6	23.0	6.3	10.5	8.3	70	2
192	10 YR 8/1	0	0	6.9	34.2	4.8	10.8	16.2	27.1	---	93	4
193	10 YR 9/1	0.9	1.5	7.7	45.6	5.0	7.4	7.4	24.5	8.6	83	2
204	10 YR 10/1	0	0	20.0	63.5	1.0	5.5	2.5	7.5	---	34	Trace
205	10 YR 8/2	0	0	3.8	69.8	0.7	10.2	4.1	11.4	---	55	26
212	10 YR 9/1	0.1	0.1	17.0	26.2	6.6	12.5	6.3	31.2	---	92	4
225	10 YR 9/1	---	15.4	23.8	12.5	8.7	18.5	7.9	13.2	---	42	24
234	10 YR 7-8/2	---	31.8	33.9	14.1	4.5	9.7	3.0	3.0	8.6	44	11
235	7.5 YR 7/5	---	21.4	41.5	16.2	4.6	10.6	1.5	4.2	---	5.3	5
236	10 YR 7/2	---	0.1	0.1	56.0	21.3	13.5	3.8	5.2	---	5.5	29
237	7.5 YR 9/1	0	0.8	4.8	40.2	11.4	25.7	6.7	10.3	---	27	45
238	10 YR 7/3	1.9	13.4	31.3	22.2	3.9	14.7	4.9	7.7	8.1	5.5	24
239	7.5 YR 6/4	0.1	0.5	5.3	7.3	3.3	37.7	17.0	28.8	8.4	26	15

Table A-1--Continued

Sample No.	Munsell Color	Texture[a]								pH	HCl Solubles (%)	Amorphous Silica (%)
240	10 YR 8/2	---	9.3	32.4	41.4	3.7	7.0	1.3	4.9	8.9	53	6
241	10 YR 8/2	---	9.7	27.0	22.1	9.5	19.8	3.9	7.9	8.3	61	20
242	10 YR 8/2	0	0.6	5.9	60.8	10.3	13.6	3.9	4.6	8.6	4.7	17
243	10 YR 8/2	0.1	2.6	10.2	59.0	1.7	13.3	5.4	7.7	---	23	22
244	7.5 YR 9/1	0	5.3	22.8	37.9	5.8	11.0	5.6	11.6	---	77	7
246	8 YR 5/5	---	12.4	37.1	24.0	7.1	14.5	1.0	3.9	8.3	10	18
247	7.5 YR 5/3	---	34.4	29.3	17.6	4.4	8.6	2.9	2.8	8.5	40	10
248	5 YR 6/7	0	0.1	0.2	36.4	21.6	24.2	5.8	11.7	8.6	44	10
249	10 YR 9/2	0	0.1	6.0	39.7	12.2	16.9	7.8	17.3	8.7	66	7
254	5 YR 6/7	0.4	7.6	6.0	64.0	7.1	9.5	1.4	4.0	---	8	12
280	7.5 YR 7/4	0.1	0.5	2.2	59.9	14.6	8.9	8.3	5.5	---	11	Trace
281	7.5 YR 5/5	0	0.3	6.0	54.4	14.2	11.4	4.9	8.8	---	---	---
282	7.5 YR 7/2	0	0.1	7.5	53.2	11.2	16.5	1.9	9.6	---	30	2
283	7.5 YR 7/2	0	0.4	5.5	51.0	10.7	17.5	5.8	9.1	---	67	Trace
284	7.5 YR 7/5	0	0.5	15.8	50.5	8.4	14.1	3.1	7.6	---	5.3	7
285	7.5 YR 7/6	0.1	2.1	19.0	42.2	9.2	14.7	3.8	8.9	---	44	1
302	10 YR 8/3	---	50.7	35.7	11.7	0.5		(1.4)		---	46	Trace
303	10 YR 8/3	0	63.5	29.8	5.6	0.4		(0.7)		---	53	Trace
304	5 YR 5/6	0.6	4.7	13.9	47.9	13.1	8.6	2.6	8.6	---	16	Trace
305	5 YR 7/3	0.1	1.4	16.7	48.6	10.3	10.7	3.8	8.4	---	43	Trace
306	5 YR 7/3	0.1	5.1	18.9	50.1	7.6	7.1	3.0	8.1	---	33	2
307	10 YR 7/3	0.1	0.1	6.5	38.4	19.5	11.4	14.6	9.6	---	2.7	Trace

308	10 YR 7/3	0	0.1	5.9	36.1	15.9	13.5	15.0	13.5	---	47	4
309	10 YR 6/3	0	0.1	1.6	40.2	20.6	16.5	12.5	8.5	---	7.2	3
310	10 YR 7/3	0	0.1	1.9	30.9	25.6	13.5	17.0	11.0	---	16	Trace
311	10 YR 6/4	1.1	4.2	5.8	24.0	18.4	9.5	24.0	13.0	---	3.3	Trace
312	5 YR 5/7	0	0.6	3.7	64.9	9.8	4.0	10.0	7.0	---	1.7	Trace
314	10 YR 6/6	0.4	1.4	4.2	30.5	14.5	9.0	24.0	16.0	---	2.3	Trace
326	10 YR 7/5	0	0.3	4.3	53.2	19.2	8.5	9.0	5.5	---	3.4	Trace
328	7.5 YR 5.5/4	0	0.2	2.9	34.1	17.8	10.0	22.0	13.0	---	4.5	2
329	10 YR 6/6	0	0.2	4.1	65.6	10.1	5.0	9.0	6.0	---	1	Trace
330	10 YR 7/4	18.8	41.1	8.7	20.2	1.7	2.5	4.5	2.5	---	1	4
331	3.5 Y 7/2	0	0.3	0.5	29.0	11.7	13.5	21.0	24.0	---	1.8	15
332	7.5 YR 8/3	3.2	23.0	17.1	30.9	6.8	2.5	5.5	7.5	---	0	Trace
333	7.5 YR 8/3	0	5.3	6.1	50.2	8.4	5.0	14.0	11.0			

[a]Textures are listed as percentages by weight in the following categories (in mm): 2 - 4.35/ 2 - .56/ .56 - .21/ .21 - .063/ .063 - .037/ .037 - .006/ .006 - .002/ <.002. Numbers in parentheses indicate the <.037 mm proportion. The fraction coarser than 2 mm is not included in the gravel matrix samples. Accuracy within the >.037 mm fraction is .3%, while in the <.037 mm fraction accuracy is thought ± 1.0%. Accuracy is low on samples where the HCl solubles and amorphous silica total over 75%.

APPENDIX B

GRAVEL ANALYSES

Data on the morphologies of clasts with major axes larger than 2 cm were aggregated following the modified Lüttig (1962) method, as outlined by Butzer (1971a, 167-169; see also Blenk, 1960; Wentworth, 1922; Krumbein, 1941). "L" refers to the length (cm) of the major axis; "I" refers to the length of the intermediate axis; and "S" refers to the length of the smallest axis. The index of rounding, "R", is the estimated percentage of convex surface on the pebble. Locations are described in Chapters 4, 5, and 8 above.

The samples were collected subjectively and judged to be representative of a particular stratum. However minor vertical and horizontal variations are common in such alluvial deposits, and these data are assumed to represent no more than a subjective average. The samples were selected from intact, primary deposits. Between 0.1 m^3 to 0.3 m^3 of deposit was passed through a 2 cm sieve, and the debris in the sieve analyzed. This analysis by individual produced a high proportion of small clasts (particularly in the larger samples) and thus the clast-size means are biased in their favor. These data are offered only for descriptive and comparative purposes, and no transport or genetic inferences are assumed. Most of the basic shapes were in fact inherited from tillites or earlier alluvial deposits (see Table 4.9 both for comparison and additional data). Gravel analysis data are presented in Table B-1.

TABLE B-1

GRAVEL ANALYSES[a]

	L	I/L	S/L	R
Droogeveld				
Total Sample--138				
$\bar{x}$	4.8	73.2	43.5	14.9
σ	1.7	13.6	14.7	16.0
CV%	36.4	18.6	33.7	107.4
Andesite Clasts Only--50				
$\bar{x}$	4.9	74.7	45.2	6.0
σ	1.9	12.6	11.6	10.2
CV%	39.2	16.4	25.7	170.0
Holpan I, Derived Gravels 1A				
Total Sample--104				
$\bar{x}$	5.1	71.1	40.2	29.3
σ	2.4	12.3	13.1	11.3
CV%	47.6	17.4	32.7	38.2
Andesite Clasts Only--66				
$\bar{x}$	5.1	70.2	37.1	28.2
σ	2.5	12.9	12.1	10.9
CV%	49.0	18.3	32.5	38.9
Holpan I, Derived Gravels 1B				
Total Sample--86				
$\bar{x}$	5.2	73.3	43.6	28.0
σ	2.1	12.3	14.2	14.8
CV%	40.6	16.8	32.7	53.0
Andesite Clasts Only--49				
$\bar{x}$	5.8	72.5	43.3	26.4
σ	2.3	12.3	14.2	15.4
CV%	39.7	16.9	38.9	58.3

TABLE B-1--Continued

	L	I/L	S/L	R
Holpan I, Alluvial Gravels				
Total Sample--117				
$\bar{x}$	5.7	71.1	37.3	30.8
σ	3.2	15.3	14.0	12.8
CV%	56.0	21.5	37.5	41.6
Andesite Clasts Only--82				
$\bar{x}$	6.4	71.0	36.9	30.7
σ	3.5	16.2	15.6	13.9
CV%	55.2	22.8	42.4	54.4
Proksch Koppie, Alluvial Gravels				
Total Sample--112				
$\bar{x}$	6.0	72.5	47.5	34.5
σ	3.7	13.3	14.6	15.6
CV%	61.3	18.4	30.6	45.3
Andesite Clasts Only--78				
$\bar{x}$	6.8	73.1	47.3	34.7
σ	4.1	13.8	14.8	16.6
CV%	60.5	18.9	31.4	47.8
Wedberg, Wedberg Terrace				
Total Sample--127				
$\bar{x}$	5.2	72.3	43.4	38.4
σ	3.4	14.5	15.4	19.4
CV%	66.0	20.1	35.4	50.4
Andesite Clasts Only--68				
$\bar{x}$	6.6	72.5	40.9	42.5
σ	3.9	14.4	15.5	19.7
CV%	59.8	19.9	37.9	46.3

TABLE B-1--Continued

	L	I/L	S/L	R
Nuwe Hoop, Wedberg Terrace				
Total Sample--75				
$\bar{x}$	5.2	72.1	27.9	27.8
σ	3.6	12.6	18.3	14.8
CV%	70.0	17.4	65.6	53.2
Shale Clasts Only--37				
$\bar{x}$	3.9	70.4	16.0	23.1
σ	1.1	11.4	12.9	14.3
CV%	27.9	16.2	80.6	61.9
Rittlemago				
Total Sample--82				
$\bar{x}$	6.2	69.3	33.9	37.2
σ	3.6	14.9	13.7	14.9
CV%	59.4	21.5	40.4	40.1
Andesite Clasts Only--44				
$\bar{x}$	6.1	67.4	35.0	37.4
σ	3.6	16.0	14.1	15.6
CV%	59.8	23.7	40.3	41.7
Riverview I, Rietputs A				
Total Sample--152				
$\bar{x}$	3.9	68.0	39.2	39.8
σ	2.0	12.5	13.0	17.3
CV%	52.0	18.4	33.2	43.5
Andesite Clasts Only--81				
$\bar{x}$	4.0	67.4	37.5	39.8
σ	2.3	13.7	13.3	18.5
CV%	57.5	20.3	35.5	46.5

TABLE B-1--Continued

	L	I/L	S/L	R
Riverview I, Rietputs B1 (one of several facies)				
Total Sample--125				
$\bar{x}$	2.5	68.8	44.3	42.5
σ	0.7	12.5	13.3	19.3
CV%	26.9	18.1	30.1	45.4
Andesite Clasts Only--60				
$\bar{x}$	2.6	70.4	45.6	40.6
σ	0.6	13.1	12.9	18.2
CV%	25.5	18.6	28.5	44.8
Riverview I, Rietputs B2				
Total Sample--120				
$\bar{x}$	4.7	69.5	44.7	39.8
σ	3.0	12.8	13.1	22.5
CV%	63.7	18.4	29.4	56.5
Andesite Clasts Only--68				
$\bar{x}$	4.7	70.5	43.6	32.1
σ	2.2	13.3	13.4	20.9
CV%	46.3	18.9	30.7	65.1
Riverview II, Rietputs A				
Total Sample--67				
$\bar{x}$	5.6	69.7	41.7	32.4
σ	3.5	13.4	12.7	13.8
CV%	62.5	19.2	30.6	42.7
Slangheuvel, weakly cemented gravel on bottom of modern channel				
Total Sample--52				
$\bar{x}$	6.0	71.9	46.1	39.6
σ	2.9	14.2	15.3	16.5
CV%	48.4	19.7	33.2	41.7

TABLE B-1--Continued

	L	I/L	S/L	R
Winter's Rush I, Fill I Gravels				
Total Sample--125				
$\bar{x}$	5.1	67.8	35.4	35.4
σ	4.0	13.1	13.9	17.7
CV%	78.8	19.3	39.3	50.2
Andesite Clasts Only--64				
$\bar{x}$	5.2	67.5	37.7	37.2
σ	3.4	13.3	13.8	19.0
CV%	66.6	19.6	36.5	51.1
Winter's Rush I, Fill II Gravels				
Total Sample--128				
$\bar{x}$	5.4	71.6	43.9	33.3
σ	3.3	13.9	14.3	17.2
CV%	60.8	19.5	32.5	51.8
Andesite Clasts Only--102				
$\bar{x}$	5.7	71.1	43.0	31.5
σ	3.5	14.1	14.8	17.1
CV%	61.4	19.8	34.4	54.3
Rietputs C at Delportshoop				
Total Sample--106				
$\bar{x}$	4.6	73.0	40.9	34.0
σ	3.4	14.4	16.6	16.2
CV%	73.2	19.7	40.7	47.7
Andesite Clasts Only--56				
$\bar{x}$	5.7	71.1	34.8	35.0
σ	4.3	15.2	17.0	18.4
CV%	75.3	21.4	49.0	52.5

TABLE B-1--Continued

	L	I/L	S/L	R
Selene, Basal Gravels				
Total Sample--80				
$\bar{x}$	6.2	70.3	28.8	34.1
σ	3.6	15.4	16.1	13.2
CV%	57.8	21.9	55.7	38.6
Douglas Gravels				
Total Sample--101				
$\bar{x}$	4.7	72.6	30.5	30.8
σ	1.8	13.5	15.2	14.8
CV%	37.9	18.6	49.8	48.2

[a] $\bar{x}$ is the mean. σ is the standard deviation. CV% is the coefficient of variation, which is $100 \cdot \sigma \cdot \bar{x}^{-1}$.

BIBLIOGRAPHY

ACOCKS, J. P. H. (1953) Veld types of the Union of South Africa. Botanical Survey of South Africa, Memoir 28, 192 pp.

ALISON, M. S. (1899) On the origin of pans. Transactions, Geological Society of South Africa, v. 4:159-161.

AMBROSE, J. W. (1964) Exhumed paleoplains of the Precambrian shield of North America. American Journal of Science, v. 262:817-857.

BAHRE, C. J. (1979) Destruction of the natural vegetation of north-central Chile. University of California Publications in Geography, v. 23, 128 pp.

BAKER, V. R. (1974) Paleohydraulic interpretation of Quaternary alluvium near Golden, Colorado. Quaternary Research, v. 4:94-112.

_______. (1977) Stream-channel response to floods, with examples from central Texas. Bulletin, Geological Society of America, v. 88:1057-1071.

BAKER, V. R., and PENTEADO-ORELLANA, M. M. (1977) Adjustment to Quaternary climatic change by the Colorado River in central Texas. Journal of Geology, v. 85:395-422.

BECK, R. (1906) Mastodon in the Pleistocene of South Africa. Geological Magazine (London), v. 5, pt. 3:49-50.

BEERBOWER, J. R. (1964) Cyclothems and cyclic depositional mechanisms in alluvial plain sedimentation. In D. F. MERRIAM (ed.), Symposium on cyclic sedimentation. State Geological Survey of Kansas, Bulletin 169, v. 1:31-42.

BIGALKE, R., and LEISTNER, O. (1972) The vegetation of Rooipoort. Unpublished manuscript, 17 pp.

BLENK, M. (1960) Ein Betrag zur morphometrischen Schotteranalyse. Zeitschrift für Geomorphologie, N.F., B. 4:202-245.

BOND, G. W. (1946) A geochemical survey of the underground water supplies of the Union of South Africa. Geological Survey, South Africa, Memoir 41, 216 pp.

_______. (1967) River valley morphology, stratigraphy, and paleoclimatology in southern Africa. In W. W. BISHOP and J. D. CLARK (eds.), Background to evolution in Africa, Chicago, University of Chicago Press, pp. 303-312.

BOSMAN, H. C. (1969) Mafeking road. Cape Town, Human and Rousseau, 166 pp.

_______. (1974) The Earth is waiting. Cape Town, Human and Rousseau, 86 pp.

BREUIL, H.; VAN RIET LOWE, C.; and DU TOIT, A. L. (1948) Early man in the Vaal River basin. Archeological Series, Archeological Survey of South Africa, v. 4, 34 pp.

BURCHELL, W. J. (1822-24) Travels in the interior of southern Africa. London, Longman, Hurst, Reese, Orme, and Brown, 2 volumes, 582 pp., 648 pp.

BUTZER, K. W. (1971a) Environment and archeology (2nd ed.). Chicago, Aldine, 703 pp.

_______. (1971b) Fine alluvial fills in the Orange and Vaal basins of South Africa. Proceedings, Association of American Geographers, v. 3:41-48.

_______. (1973) On the geology of a late Pliocene Mammuthus site, Virginia, Orange Free State. Navorsinge van die Nasionale Museum, Bloemfontein, v. 2, pt. 2:386-393.

_______. (1974a) Towards a paleo-ecology of early man in Africa: The Taung australopithecine revisited. Current Anthropology, v. 15:367-382, 413-416, 421-426.

_______. (1974b) Geo-archeological interpretation of Acheulian calc-pan sites at Doornlaagte and Rooidam (Kimberley, South Africa). Journal of Archeological Science, v. 1:1-25.

_______. (1976) Lithostratigraphy of the Swartkrans Formation. South African Journal of Science, v. 72:136-141.

BUTZER, K. W.; FOCK, G. J.; STUCKENRATH, R.; and ZILCH, A. (1973) Paleohydrology of late-Pleistocene Lake Alexandersfontein, Kimberley, South Africa. Nature, v. 243:328-330.

BUTZER, K. W.; FOCK, G. J.; SCOTT, L.; and STUCKENRATH, R. (1979) Dating and context of rock engravings in southern Africa. Science, v. 203:1201-1214.

BUTZER, K. W., and HELGREN, D. M. (1972) Late-Cenozoic evolution of the Cape coast between Knysna and Cape St. Francis, South Africa. Quaternary Research, v. 2:143-169.

BUTZER, K. W.; HELGREN, D. M.; FOCK, G. J.; and STUCKENRATH, R. (1973) Alluvial terraces of the lower Vaal River, South Africa: A reappraisal and reinvestigation. Journal of Geology, v. 81:341-362.

BUTZER, K. W., and ISAAC, G. L. (1975) After the australopithecines. The Hague, Mouton, 911 pp.

BUTZER, K. W.; STUCKENRATH, R.; BRUZEWICZ, A. J.; and HELGREN, D.M. (1978) Late Cenozoic paleoclimates of the Gaap Escarpment, Kalahari Margin, South Africa. Quaternary Research, v. 10:514-543.

CAMPBELL, J. (1816) Travels in South Africa. Andover (Mass.), Flagg and Gould, 398 pp.

CAMPBELL, J. (1822) Travels in South Africa, undertaken at the request of the London Missionary Society; being a narrative of a second journey in the interior of that country. London, London Missionary Society, 2 volumes, 322 pp., 384 pp.

CARSON, M. A. and KIRKBY, M.S. (1972) Hillslope form and process. London, Cambridge University Press, 475 pp.

CHOW, V. T. (1959) Open channel hydraulics. New York, McGraw-Hill, 680 pp.

CHURCH, M. (1972) Baffin Island sandurs: A study of arctic fluvial processes. Geological Survey of Canada, Bulletin 216, 208 pp.

CHURCH, M., and GILBERT, R. (1975) Proglacial fluvial and lacustrine environments. In A. V. JOPLING and B. C. Mc DONALD (eds.), Glaciofluvial and glaciolacustrine sedimentation. Society of Economic Paleontologists and Mineralogists, Special Publication No. 23, pp. 22-100.

CLAGUE, J. J. (1975) Sedimentology and paleohydrology of late Wisconsinan outwash, Rocky Mountain Trench, southeastern British Columbia. In A. V. JOPLING and B. C. Mc DONALD (eds.), Glaciofluvial and glaciolacustrine sedimentation. Society of Economic Paleontologists and Mineralogists, Special Publication No. 23, pp. 223-237.

COLE, G. H. (1961) Culture change in the middle-upper Pleistocene transition in Africa south of the Sahara. Unpublished Ph.D. dissertation, University of Chicago, 364 pp.

COOKE, H. B. S. (1947) The development of the Vaal River and its deposits. Transactions, Geological Society of South Africa, v. 46:243-260.

_______. (1949) Fossil mammals of the Vaal River deposits. Geological Survey, South Africa. Memoir 35, Part 3, 109 pp.

_______. (1963) Pleistocene mammal faunas of Africa, with particular reference to southern Africa. In F. C. HOWELL and F. BOURLIERE (eds.), African ecology and human evolution, Chicago, Aldine Press, pp. 65-116.

_______. (1967) The Pleistocene sequence in southern Africa and problems of correlation. In W. W. BISHOP and J. D. CLARK (eds.), Background to evolution in Africa, Chicago, University of Chicago Press, pp. 175-184.

COOKE, H. B. S., and MAGLIO, V. J. (1972) Plio-Pleistocene stratigraphy in East Africa in relation to proboscidean and suid evolution. In W. W. BISHOP and J. A. MILLER (eds.), Calibration of Hominoid evolution, Edinburgh, Scottish Academic Press, pp. 303-329.

COOKE, R. U., and REEVES, R. W. (1976) Arroyos and environmental change in the American South-West. Oxford, Clarendon Press, 213 pp.

COPPENS, Y.; HOWELL, F. C.; ISAAC, G. Ll,; and LEAKEY, R. E. F. (eds.) (1976) Earliest man and environments in the Lake Rudolf basin: stratigraphy, paleoecology, and evolution. Chicago, University of Chicago Press, 615 pp.

COTTER, E. (1971) Paleoflow characteristics of a late-Cretaceous river in Utah from analysis of sedimentary structures in the Ferron Sandstone. Journal of Sedimentary Petrology, v. 41:129-138.

DART, R. A. (1925) Australopithecus africanus, the man-ape of South Africa. Nature, v. 115:195-199.

_______. (1929) Mammoths and other fossil elephants of the Vaal and Limpopo watersheds. South African Journal of Science, v. 26:698-731.

DAVIS, S. N. (1958) Size distribution of rock fragments in stream gravel and glacial till. Journal of Sedimentary Petrology, v. 28:87-94.

_______. (1964) Silica in streams and groundwater. American Journal of Science, v. 262, p. 870-891.

DAVIS, W. M. (1906) Observations in South Africa. Bulletin, Geological Society of America, v. 17:337-450.

DEPARTMENT OF WATER AFFAIRS (1973) Stream gauge records along the lower Riet, Modder, Vaal, and Harts rivers, and the middle Orange River. Unpublished photocopies.

DE SWARDT, A. M. J. (1974) Geomorphological dating of cave openings in South Africa. Nature, v. 250:683.

DE SWARDT, A. M. J., and BENNET, G. (1974) Structural and physiographic development of Natal since the late Jurassic. Transactions, Geological Society of South Africa, v. 77:309-322.

DINGLE, R. V. (1973) The geology of the continental shelf between Luderitz and Cape Town (South West Africa), with special reference to Tertiary strata. Journal, Geological Society of London, v. 129:337-363.

DIXEY, F. (1938) Some observations on the physiographical development of central and southern Africa. Transactions, Geological Society of South Africa, v. 41:113-170.

_______. (1942) Erosion cycles in central and southern Africa. Transactions, Geological Society of South Africa, v. 45:151-167.

DURY, G. H. (1974) Magnitude-frequency analysis and channel morphometry. In M. MORISAWA (ed.), Fluvial Geomorphology, State University Publications in Geomorphology, pp. 91-121.

_______. (1976) Discharge prediction, present and former, from channel dimensions. Journal of Hydrology, v. 30:219-245.

DU TOIT, A. L. (1906) Geological survey of portions of the divisions of Vryburg and Mafeking. Annual Report for 1905 of the Geological Commission of the Cape of Good Hope, pp. 207-258.

_______. (1907) Geological survey of the eastern portion of Griqualand West. Annual Report for 1906 of the Geological Commission of the Cape of Good Hope, pp. 87-176.

_______. (1908) Geological survey of portions of Hopetown, Britstown, Prieska, and Hay. Annual Report for 1907 of the Geological Commission of the Cape of Good Hope, pp. 161-192.

_______. (1910) The evolution of the river system of Griqualand West. Transactions, Royal Society of South Africa, v. 1:347-362.

_______. (1933) Crustal movement as a factor in the geological evolution of South Africa. South African Geographical Journal, v. 16:3-20.

DU TOIT, A. L. (1951) The diamondiferous gravels of Lichtenburg. Geological Survey, South Africa, Memoir 44, 50 pp.

_______. (1953) The geology of South Africa (3rd ed., S. H. HAUGHTON, ed.). New York, Hafner, 611 pp.

DU TOIT, S. R., and LEITH, M. J. (1974) The J(c)-1 bore-hole on the continental shelf near Stanger, Natal. Transactions, Geological Society of South Africa, v. 77:247-252.

EMBLETON, C., and KING, C. A. M. (1968) Glacial and periglacial geomorphology. New York, St. Martins Press, 608 pp.

EMERY, K. O. (1972) Eastern Atlantic continental margin: some results of the 1972 cruise of R. V. Atlantis II. Science, v. 178:298-300.

ENBELBRECHT, L. N. J. (1973) Die geologie van die gebied tussen Kimberley en Barkly-Wes, Kaapprovinsie. Unpublished M.Sc. thesis, University of the Orange Free State, 105 pp.

FAHNESTOCK, R. K. (1963) Morphology and hydrology of a glacial stream -- White River, Mount Ranier, Washington. U.S. Geological Survey Professional Paper 422-A, 70 pp.

FAIR, T. J. D. (1948) Hillslopes and pediments in the semi-arid Karroo. South African Geographical Journal, v. 30:71-79.

FAIRBRIDGE, R. W. (1968) Fluvial terraces. In R. W. FAIRBRIDGE (ed.), Encyclopedia of Geomorphology. New York, Reinhold, pp. 1117-1140.

FITCH, F. J., and MILLER, J. A. (1971) Potassium-argon radioages of Karroo volcanic rocks in Lesotho. Bulletin Volcanologique, v. 35:64-84.

FLINT, R. F. (1959) Pleistocene climates in eastern and southern Africa. Bulletin, Geological Society of America, v. 70:343-374.

_______. (1971) Glacial and Quaternary geology. New York, John Wiley, 892 pp.

FLINT, R. F., and BOND, G. (1968) Pleistocene sand ridges and pans in western Rhodesia. Bulletin, Geological Society of America, v. 79:299-314.

FOCK, G. J. (1960) Die Prä-Chelles-Kultur am Vaal. In Steinzeitfragen der Alten und Neuen Welt. Festschrift für L. Zotz, Bonn, Röhrscheid, pp. 151-160.

FRYE, J. C. (1958) Climate and Lester King's "Uniformitarian Nature of Hillslopes." Journal of Geology, v. 67:111-113.

GARNER, H. F. (1974) The origin of landscapes. New York, Oxford University Press, 734 pp.

GEYSER, G. W. P. (1950) Panne--hul ontstaan en die faktore wat daartoe aanleiding gee. South African Geographical Journal, v. 32:15-31.

GOODWIN, A. J. H. (1928) Archeology of the Vaal River gravels. Transactions, Royal Society of South Africa, v. 16:77-102.

GOUDIE, A. S. (1972) The chemistry of world calcrete. Journal of Geology, v. 80:449-463.

_______. (1973) Duricrusts in tropical and subtropical landscapes. Oxford, Clarendon Press, 174 pp.

GRAVENOR, C. P. (1975) Erosion by continental ice sheets. American Journal of Science, v. 275:594-604.

GREGORY, K. J., and WALLING, D. E. (1973) Drainage basin form and process. New York, John Wiley, 456 pp.

HAMILTON, W., and KRINSLEY, D. (1967) Upper Paleozoic glacial deposits of South Africa and southern Australia. Bulletin, Geological Society of America, v. 78:783-800.

HARLAND, W. B.; HEROD, K. N.; and KRINSLEY, D. (1966) The definition and identification of tills and tillites. Earth Science Reviews, v. 2:225-256.

HARRIS, J. M. (1976) Pliocene giraffoidea (mammalia, artiodactyla) from Cape Province. Annals, South African Museum, v. 69:325-353.

HARRIS, W. C. (1839) The wild sports of southern Africa ... . London, J. Murray, 387 pp.

HAUGHTON, S. H. (1969) Geological history of southern Africa. Cape Town, Geological Society of South Africa, 535 pp.

HAWTHORNE, J. B. (1975) Model of a kimberlite pipe. Physics and Chemistry of the Earth, v. 9:1-15.

HELGREN, D. M. (1977) Geological context of the Vaal River faunas. South African Journal of Science, v. 73:303-307.

_______. (1978) Acheulian settlement along the lower Vaal River, South Africa. Journal of Archeological Science, v. 5:39-60.

HELGREN, D. M., and BUTZER, K. W. (1974) Alluvial terraces of the lower Vaal River, South Africa: A reappraisal and reinvestigation: A reply. Journal of Geology, v. 82:665-667.

_______. (1977) Paleosols of the southern Cape coast, South Africa: Implications for laterite definition, genesis, and age. Geographical Review, v. 67:430-445.

HELLEY, E. J. (1969) Field measurement of the initiation of large bed particle motion in Blue Creek near Klamath, California. U.S. Geological Survey Professional Paper 562-G, 19 pp.

HJULSTRÖM, F. (1935) Studies of the morphological activity of rivers as illustrated by the river Fyris. Bulletin, Geological Institute, Uppsala, v. 25:221-527.

HOLUB, E. (1881) Seven years in South Africa: Travels, researches, hunting adventures between the diamond fields and the Zambesi (1872-79) (2nd ed.). London, Sampson, Low, Marston, Searle, and Rivington, 2 volumes, 426 pp., 479 pp.

HUMPHREYS, A. J. B. (1969) Later Acheulian or Fauresmith? A contribution. Annals, Cape Provincial Museums, v. 6, pt..10:87-101.

JOPLING, A. V. (1966) Some principles and techniques used in reconstructing the hydraulic parameters of a paleo-flow regime. Journal of Sedimentary Petrology, v. 36:5-49.

KENT, L. E., and ROGERS, A. W. (1947) Diatomaceous deposits in the Union of South Africa, with special reference to the Kieselguhr. Geological Survey, South Africa, Memoir 42, 260 pp.

KING, L. C. (1963) South African scenery, a textbook of geomorphology, (3rd ed., revised). Edinburgh, Oliver and Boyd, 308 pp.

_______. (1967) The morphology of the earth -- a study and synthesis of world scenery, (2nd ed.). New York, Hafner, 726 pp.

_______. (1972) The Natal monocline: explaining the origin and scenery of Natal, South Africa. Durban, University of Natal, Geology Department, 113 pp.

_______. (1976) Comment upon two papers concerning the geomorphology of Natal. Transactions, Geological Society of South Africa, v. 79: 149-151.

KRAUSKOPF, K. B. (1959) The geochemistry of silica in sedimentary environments. In H. A. IRELAND (ed.), Silica in sediments. Society of Economic Paleontologists and Mineralogists Special Publication No. 7, pp. 4-19.

KRUMBEIN, W. C. (1941) The effects of abrasion on the size, shape, and roundness of rock fragments. Journal of Geology, v. 49:482-520.

LINK, A. G. (1966) Textural classification of sediments. Sedimentology, v. 7:249-254.

LUTTIG, G. (1962) The shape of pebbles in continental, fluviatile, and marine facies. International Association for Scientific Hydrology, Pub. 59, pp. 253-258.

MAARLEVELD, G. C. (1960) Über die pleistozänen Ablagerungen im südlichen Afrika. Erdkunde, v. 14:35-46.

MAC GREGOR, I. M.; TRUSWELL, J. F.; and ERIKSSON, K. A. (1974) Filamentous algae from the 2300 m.y. old Transvaal dolomite. Nature, v. 247:538-539.

MAGLIO, V. J. (1973) Origin and evolution of the Elephantidae. Transactions, American Philosophical Society, v. 63(n.s.), pt. 3:1-149.

MALAN, B. D. (1947) Flake tools and artefacts in the Stellenbosch-Fauresmith transition in the Vaal River valley. South African Journal of Science, v. 43:350-362.

MARTIN, L. (1916) The physical geography of Wisconsin. Bulletin, Wisconsin Geological and Natural History Survey, v. 36:1-549.

MASON, R. J. (1962) Prehistory of the Transvaal: A record of human activity. Johannesburg, University of the Witwatersrand Press, 498 pp.

_______. (1967) The archeology of the earliest surficial deposits in the lower Vaal basin near Holpan, Windsorton District. South African Geographical Journal, v. 49:39-56.

MAVIS, E. T., and LAUSHEY, L. M. (1949) Formula for velocity of beginning of bedload movement is reappraised. Civil Engineering, v. 19:38-39.

MAYER, J. J. (1973) Morphotectonic development of the Harts River valley in relation to the Griqualand-Transvaal axis and the Vaal and Molopo Rivers. Transactions, Geological Society of South Africa, v. 76:183-194.

MC DOUGALL, I. (1963) Potassium-argon age measurements on dolerites from Antarctica and South Africa. Journal of Geophysical Research, v. 68:1535-45.

MC LACHLAN, I. R., and ANDERSON, A. (1973) A review of the evidence for marine conditions in southern Africa during Dwyka times. Paleontologica Africana, v. 15:37-64.

MERCER, J. H. (1976) Glacial history of southernmost South America. Quaternary Research, v. 6:125-166.

MOLLARD, J. D. (1957) A study of aerial mosaics in southern Saskatchewan and Manitoba. Oil in Canada, August 5, 1957.

MORISAWA, M. (1968) Streams. New York, McGraw-Hill, 175 pp.

MUNSELL COLOR COMPANY (1954) Munsell soil color charts. Baltimore, Munsell Soil Color Company.

NETTERBERG, F. (1969a) Ages of calcretes in southern Africa. South African Archeological Bulletin, v. 24:88-92.

NETTERBERG, F. (1969b) The interpretation of some basic calcrete types. South African Archeological Bulletin, v. 24:117-122.

_______. (1974) Observations on calcretes at some archeological and paleontological sites in South Africa. Goodwin Series, South African Archeological Society, v. 2:20-24.

PARTRIDGE, T. C. (1969) Fluvial features and climatic change during the Quaternary in South Africa. South African Archeological Bulletin, v. 24:106-116.

PARTRIDGE, T. C., and BRINK, A. B. A. (1967) Gravels and terraces of the lower Vaal basin. South African Geographical Journal, v. 49:21-38.

_______. (1974) Alluvial terraces of the lower Vaal River, South Africa: A reappraisal and reinvestigation: A discussion. Journal of Geology, v. 82:663-665.

PEABODY, F. E. (1954) Travertines and cave deposits of the Kaap Escarpment, South Africa. Bulletin, Geological Society of America, v. 65: 671-706.

PIAGET, J. E. H. (1963) Coarse fraction mineralogy and granulometry of selected soils of the western Orange Free State. Unpublished D. Sc. thesis, University of the Orange Free State, 255 pp.

POWERS, H. C. (1953) A new roundness scale for sedimentary particles. Journal of Sedimentary Petrology, v. 23:117-119.

PRESTON-WHYTE, R. A. (1974) Climatic classification of South Africa: A multivariate approach. South African Geographical Journal, v. 56:79-86.

PRESTON-WHYTE, R. A., and WHITTINGTON, R. F. (1972) Mean rainfall and mean runoff in South Africa: An investigation into phase differences. South African Geographical Journal, v. 54:133-137.

READING, H. G., and WALKER, R. G. (1966) Sedimentation of Eocambrian tillites and associated sediments in Finnmark, northern Norway. Paleogeography, Paleoclimatology, and Paleoecology, v. 2:177-212.

ROGERS, A. W. (1907) Geological survey of parts of Bechuanaland and Griqualand West. Report for 1906 of the Geological Commission of the Cape of Good Hope, pp. 7-85.

ROGERS, A. W., and DU TOIT, A. L. (1909) An introduction to the geology of Cape Colony. London, Longmans, Green, and Company, 491 pp.

ROGERS, A. W., and SCHWARZ, E. H. L. (1900) Geology of the Orange River valley in the Hopetown and Prieska districts. Annual Report for 1899 of the Geological Commission of the Cape of Good Hope, pp. 65-97.

ROWSELL, D. M., and DE SWARDT, A. M. J. (1976) Diagenesis in Cape and Karroo sediments, South Africa, and its bearing on their hydrocarbon potential. Transactions, Geological Society of South Africa, 79:81-145.

RUST, B. R. (1972) Structure and process in a braided river. Sedimentology, v. 18:221-245.

SAMPSON, C. G. (1974) The stone age archeology of southern Africa. New York, Academic Press, 518 pp.

SCHULZE, B. R. (1965) General survey: climate of South Africa, part 8. Pretoria, South African Weather Bureau, 330 pp.

SCHUMM, S. A. (1960) The shape of alluvial channels in relation to sediment type. U.S. Geological Survey Professional Paper 352-B, pp. 17-30.

_______. (1963) Sinuosity of alluvial channels on the Great Plains. Bulletin, Geological Society of America, v. 74:1089-1100.

_______. (1968) River adjustment to altered hydrologic regimen -- Murrumbidgee River and paleochannels, Australia. U.S. Geological Survey Professional Paper 598, 65 pp.

_______. (1972) Fluvial paleochannels. In J. K. RIGBY and W. K. HAMBLIN (eds.), Recognition of ancient sedimentary environments. Society of Economic Paleontologists and Mineralogists, Special Publication No. 16, pp. 98-107.

_______. (1973) Geomorphic thresholds and complex response of drainage systems. In M. MORISAWA (ed.), Fluvial geomorphology. New York, State University Publications in Geomorphology, pp. pp. 299-310.

_______. (1977) The Fluvial System. New York, John Wiley, 338 pp.

SCHUMM, S. A., and LICHTY, R. A. (1963) Channel widening and floodplain construction along Cimarron River in southwestern Kansas. U.S. Geological Survey Professional Paper 352-D, pp. 71-88.

SCOTT, L. (1976) Preliminary palynological results from the Alexandersfontein basin near Kimberley. Annals, South African Museum, v. 71:193-199.

SHACKLETON, N. J., and KENNETT, J. P. (1975) Late Cenozoic oxygen and carbon isotopic ages at DSDP site 284: Implications for glacial history of the Northern Hemisphere and Antarctica. Initial Reports of the Deep Sea Drilling Project, v. 29:801-807.

SHEPHERD, R. G., and SCHUMM, S. A. (1974) Experimental study of river incision. Bulletin, Geological Society of America, v. 85:257-268.

SIEVER, R. (1957) The silica budget in the sedimentary cycle. American Mineralogist, v. 42:821-841.

SÖHNGE, P. G., and VISSER, D. J. L. (1937) The geology of the Vaal River basin. Geological Survey, South Africa, Memoir 35, part 1, pp. 5, 7-59, 165-176, plates I-VI.

SOIL SURVEY STAFF (1975) Soil Taxonomy. Soil Conservation Service, U.S. Department of Agriculture, Agriculture Handbook No. 436, 754. pp.

SOUTH AFRICAN WEATHER BUREAU (1954) Climate statistics: Climate of South Africa, part 1. Pretoria, South African Weather Bureau, 160 pp.

________. (1960) Surface winds: Climate of South Africa, part 6. Pretoria, South African Weather Bureau, 202 pp.

_______. (1965) Average monthly rainfall up to the end of 1960: Climate of South Africa, part 9. Pretoria, South African Weather Bureau, 359 pp.

STEEDMAN, A. (1835) Wanderings and adventures in the interior of southern Africa. London, Longman and Company, 2 volumes, 330 pp., 358 pp.

STOW, G. W. (1872) On the diamond gravels of the Vaal River, South Africa. Quaterly Journal, Geological Society of London, v. 28:3-21.

_______. (1874) Geological notes upon Griqualand West. Quarterly Journal, Geological Society of London, v. 30:581-680.

STRATTEN, T. (1968) The Dwyka glaciation and its relationship to the pre-Karroo surface. Unpublished D.Sc. thesis, University of the Witwatersrand, 2 volumes, 196 pp., 100 pp.

STRATTEN, T., and HUMPHREYS, A. J. B. (1974) Extensive glacial pavement of Dwyka age near Douglas, Cape Province. South African Journal of Science, v. 70:44-45.

SZABO, B. J. and BUTZER, K. W. (in press) Uranium-series dating of lacustine limestones from pan deposits with final Acheulian assemblage at Rooidan, Kimberley District, South Africa.

THOMPSON, G. (1827) Travels and adventures in southern Africa. London Henry Colburn, 493 pp.

TOENS, P. D. (1966) Precambrian dolomite and limestone of the northern Cape Province. Geological Survey, South Africa, Memoir 57, 109 pp.

TROLLOPE, A. (1878) South Africa. London, Chapman and Hall, 2 volumes, 352 pp., 352 pp.

VAN DER EYK, J. J.; MAC IVOR, C. N.; and DE VILLIERS, J. M. (1969) Soils of the Tugela Basin -- A study in sub-tropical Africa. Pietermaritzburg, The (Natal) Town and Regional Planning Commission.

VAN EEDEN, O. R. (1955) Die ontsaan van panne in Suid-Afrika. Tegnikon, July (1955), pp. 206-215.

_______. (1963) The geology of the area around Schweizer-Reneke. Geological Survey, South Africa, explanation to sheets 2724B and 2725A.

VAN NIEKERK, C. B., and BURGER, A. J. (1964) The age of the Ventersdorp System. Annals, Geological Survey of South Africa, v. 3:75-86.

VAN RIET LOWE, C. (1935) Implementiferous gravels of the Vaal River at Riverview Estates. Nature, v. 136:53-55.

_______. (1937) The archeology of the Vaal River basin. Geological Survey, South Africa, Memoir 35, Part 2, pp. 6, 7, 61-164, 177-184.

_______. (1952) The Vaal River chronology -- an up-to-date summary. South African Archeological Bulletin, v. 7(28):135-149.

VAN ROOYEN, T. H. (1972) Soils of the central Orange River basin. Unpublished D.Sc. thesis, University of the Orange Free State.

VAN ROOYEN, T. H., and BURGER, R. DU T. (1974a) Plant-ecological significance of the soils of the central Orange River basin. South African Geographical Journal, v. 56:60-66.

_______. (1974b) Dispersal and sedimentary characteristics of sandy soil parent materials in semiarid South Africa. Zeitschrift für Geomorphologie, Supplemental Band 20:69-90.

VAN VREEDEN, B. F. (1961) Die oorsprong en geskiedenis van plekname in Noord-Kaapland en die aangrensende gebiede. Unpublished Ph.D. dissertation, University of Witwatersrand, 707 pp.

VAN ZINDERN BAKKER, E. M., and BUTZER, K. W. (1974) Quaternary environmental changes in southern Africa. Soil Science, v. 116:236-248.

VISSER, J. N. J., and GROBLER, N. J. (1972) The transition beds at the base of the dolomite series in the northern Cape Province. Transactions, Geological Society of South Africa, v. 75:265-274.

VONNEGUT, K. (1969) Slaughterhouse-Five or The Children's Crusade. Tralfamadore, Delacorte Press, 186 pp.

VRBA, E. S. (1975) Some evidence of chronology and paleo-ecology of Sterkfontein, Swartkrans, and Kromdraai from the fossil bovidae. Nature, v. 254:301-304.

WAGNER, P. A. (1914) The diamond fields of South Africa. Johannesburg, The Transvaal Leader, 347 pp.

WALKER, F., and POLDERVAART, A. (1949) Karroo dolerites of the Union of South Africa. Bulletin, Geological Society of America, v. 60:591-706.

WELLINGTON, J. H. (1937) The pre-Karroo peneplain in the south-central Transvaal. South African Journal of Science, v. 33:281-295.

_______. (1943) The Lake Chrissie problem. South African Geographical Journal, v. 25:50-64.

_______. (1945) Notes on the drainage of the western Free State sandveld. South African Geographical Journal, v. 27:73-77.

_______. (1955) Southern Africa -- a geographical study, volume 1, physical geography. Cambridge, University Press, 528 pp.

WELLS, L. H. (1964) The Vaal River Younger Gravels faunal assemblage: A revised list. South African Journal of Science, v. 60:91-93.

WENTWORTH, C. K. (1922) A field study of the shapes of pebbles. U.S. Geological Survey, Bulletin 730, pp. 81-114.

WILLIAMS, A. F. (1932) The genesis of the diamond. London, Bouverie House, 636 pp.

WILLIAMS, P. F., and RUST, B. R. (1969) The sedimentary petrology of a braided river. Journal of Sedimentary Petrology, v. 39:649-679.

WILMAN, M. (1946) Preliminary checklist of the flowering plants and ferns of Griqualand West (Southern Africa). Kimberley, Alexander McGregor Memorial Museum, 381 pp.

WINTER, H. DE LA R. (1976) A lithostratigraphic classification of the Ventersdorp succession. Transactions, Geological Society of South Africa, v. 79:31-48.

WOLMAN, M. G., and MILLER, J.P. (1960) Magnitude and frequency of forces in geomorphic processes. Journal of Geology, v. v. 68:54-74.

YOUNG, A. (1972) Slopes. Edinburgh, Oliver and Boyd, 288 pp.

ZEUNER, F. E. (1945) The Pleistocene Period, 1st ed. (2nd ed., 1959), London, Hutchinson, 447 pp.

THE UNIVERSITY OF CHICAGO

DEPARTMENT OF GEOGRAPHY

RESEARCH PAPERS (Lithographed, 6×9 Inches)

(*Available from Department of Geography, The University of Chicago, 5828 S. University Ave., Chicago, Illinois 60637. Price: $8.00 each; by series subscription, $6.00 each.*)

106. SAARINEN, THOMAS F. *Perception of the Drought Hazard on the Great Plains* 1966. 183 pp.
107. SOLZMAN, DAVID M. *Waterway Industrial Sites: A Chicago Case Study* 1967. 138 pp.
108. KASPERSON, ROGER E. *The Dodecanese: Diversity and Unity in Island Politics* 1967. 184 pp.
109. LOWENTHAL, DAVID, et al. *Environmental Perception and Behavior.* 1967. 88 pp.
110. REED, WALLACE E. *Areal Interaction in India: Commodity Flows of the Bengal-Bihar Industrial Area* 1967. 210 pp.
112. BOURNE, LARRY S. *Private Redevelopment of the Central City: Spatial Processes of Structural Change in the City of Toronto* 1967. 199 pp.
113. BRUSH, JOHN E., and GAUTHIER, HOWARD L., JR. *Service Centers and Consumer Trips: Studies on the Philadelphia Metropolitan Fringe* 1968. 182 pp.
114. CLARKSON, JAMES D. *The Cultural Ecology of a Chinese Village: Cameron Highlands, Malaysia* 1968. 174 pp.
115. BURTON, IAN; KATES, ROBERT W.; and SNEAD, RODMAN E. *The Human Ecology of Coastal Flood Hazard in Megalopolis* 1968. 196 pp.
117. WONG, SHUE TUCK. *Perception of Choice and Factors Affecting Industrial Water Supply Decisions in Northeastern Illinois* 1968. 96 pp.
118. JOHNSON, DOUGLAS L. *The Nature of Nomadism* 1969. 200 pp.
119. DIENES, LESLIE. *Locational Factors and Locational Developments in the Soviet Chemical Industry* 1969. 285 pp.
120. MIHELIC, DUSAN. *The Political Element in the Port Geography of Trieste* 1969. 104 pp.
121. BAUMANN, DUANE. *The Recreational Use of Domestic Water Supply Reservoirs: Perception and Choice* 1969. 125 pp.
122. LIND, AULIS O. *Coastal Landforms of Cat Island, Bahamas: A Study of Holocene Accretionary Topography and Sea-Level Change* 1969. 156 pp.
123. WHITNEY, JOSEPH. *China: Area, Administration and Nation Building* 1970. 198 pp.
124. EARICKSON, ROBERT. *The Spatial Behavior of Hospital Patients: A Behavioral Approach to Spatial Interaction in Metropolitan Chicago* 1970. 198 pp.
125. DAY, JOHN C. *Managing the Lower Rio Grande: An Experience in International River Development* 1970. 277 pp.
126. MAC IVER, IAN. *Urban Water Supply Alternatives: Perception and Choice in the Grand Basin, Ontario* 1970. 178 pp.
127. GOHEEN, PETER G. *Victorian Toronto, 1850 to 1900: Pattern and Process of Growth* 1970. 278 pp.
128. GOOD, CHARLES M. *Rural Markets and Trade in East Africa* 1970. 252 pp.
129. MEYER, DAVID R. *Spatial Variation of Black Urban Households* 1970. 127 pp.
130. GLADFELTER, BRUCE. *Meseta and Campiña Landforms in Central Spain: A Geomorphology of the Alto Henares Basin* 1971. 204 pp.
131. NEILS, ELAINE M. *Reservation to City: Indian Urbanization and Federal Relocation* 1971. 200 pp.
132. MOLINE, NORMAN T. *Mobility and the Small Town, 1900–1930* 1971. 169 pp.
133. SCHWIND, PAUL J. *Migration and Regional Development in the United States, 1950–1960* 1971. 170 pp.
134. PYLE, GERALD F. *Heart Disease, Cancer and Stroke in Chicago: A Geographical Analysis with Facilities Plans for 1980* 1971. 292 pp.
135. JOHNSON, JAMES F. *Renovated Waste Water: An Alternative Source of Municipal Water Supply in the U.S.* 1971. 155 pp.
136. BUTZER, KARL W. *Recent History of an Ethiopian Delta: The Omo River and the Level of Lake Rudolf* 1971. 184 pp.
137. HARRIS, CHAUNCY D. *Annotated World List of Selected Current Geographical Serials in English, French, and German* 3rd edition 1971. 77 pp.
138. HARRIS, CHAUNCY D., and FELLMANN, JEROME D. *International List of Geographical Serials* 2nd edition 1971. 267 pp.
139. MC MANIS, DOUGLAS R. *European Impressions of the New England Coast, 1497–1620* 1972. 147 pp.
140. COHEN, YEHOSHUA S. *Diffusion of an Innovation in an Urban System: The Spread of Planned Regional Shopping Centers in the United States, 1949–1968* 1972. 136 pp.

141. MITCHELL, NORA. *The Indian Hill-Station: Kodaikanal* 1972. 199 pp.

142. PLATT, RUTHERFORD H. *The Open Space Decision Process: Spatial Allocation of Costs and Benefits* 1972. 189 pp.

143. GOLANT, STEPHEN M. *The Residential Location and Spatial Behavior of the Elderly: A Canadian Example* 1972. 226 pp.

144. PANNELL, CLIFTON W. *T'ai-chung, T'ai-wan: Structure and Function* 1973. 200 pp.

145. LANKFORD, PHILIP M. *Regional Incomes in the United States, 1929–1967: Level, Distribution, Stability, and Growth* 1972. 137 pp.

146. FREEMAN, DONALD B. *International Trade, Migration, and Capital Flows: A Quantitative Analysis of Spatial Economic Interaction* 1973. 202 pp.

147. MYERS, SARAH K. *Language Shift Among Migrants to Lima, Peru* 1973. 204 pp.

148. JOHNSON, DOUGLAS L. *Jabal al-Akhdar, Cyrenaica: An Historical Geography of Settlement and Livelihood* 1973. 240 pp.

149. YEUNG, YUE-MAN. *National Development Policy and Urban Transformation in Singapore: A Study of Public Housing and the Marketing System* 1973. 204 pp.

150. HALL, FRED L. *Location Criteria for High Schools: Student Transportation and Racial Integration* 1973. 156 pp.

151. ROSENBERG, TERRY J. *Residence, Employment, and Mobility of Puerto Ricans in New York City* 1974. 230 pp.

152. MIKESELL, MARVIN W., editor. *Geographers Abroad: Essays on the Problems and Prospects of Research in Foreign Areas* 1973. 296 pp.

153. OSBORN, JAMES. *Area, Development Policy, and the Middle City in Malaysia* 1974. 273 pp.

154. WACHT, WALTER F. *The Domestic Air Transportation Network of the United States* 1974. 98 pp.

155. BERRY, BRIAN J. L., et al. *Land Use, Urban Form and Environmental Quality* 1974. 464 pp.

156. MITCHELL, JAMES K. *Community Response to Coastal Erosion: Individual and Collective Adjustments to Hazard on the Atlantic Shore* 1974. 209 pp.

157. COOK, GILLIAN P. *Spatial Dynamics of Business Growth in the Witwatersrand* 1975. 143 pp.

158. STARR, JOHN T., JR. *The Evolution of Unit Train Operations in the United States: 1960–1969—A Decade of Experience* 1976. 247 pp.

159. PYLE, GERALD F. *The Spatial Dynamics of Crime* 1974. 220 pp.

160. MEYER, JUDITH W. *Diffusion of an American Montessori Education* 1975. 109 pp.

161. SCHMID, JAMES A. *Urban Vegetation: A Review and Chicago Case Study* 1975. 280 pp.

162. LAMB, RICHARD. *Metropolitan Impacts on Rural America* 1975. 210 pp.

163. FEDOR, THOMAS. *Patterns of Urban Growth in the Russian Empire during the Nineteenth Century* 1975. 275 pp.

164. HARRIS, CHAUNCY D. *Guide to Geographical Bibliographies and Reference Works in Russian or on the Soviet Union* 1975. 496 pp.

165. JONES, DONALD W. *Migration and Urban Unemployment in Dualistic Economic Development* 1975. 186 pp.

166. BEDNARZ, ROBERT S. *The Effect of Air Pollution on Property Value in Chicago* 1975. 118 pp.

167. HANNEMANN, MANFRED. *The Diffusion of the Reformation in Southwestern Germany, 1518-1534* 1975. 248 pp.

168. SUBLETT, MICHAEL D. *Farmers on the Road. Interfarm Migration and the Farming of Noncontiguous Lands in Three Midwestern Townships, 1939-1969* 1975. 228 pp.

169. STETZER, DONALD FOSTER. *Special Districts in Cook County: Toward a Geography of Local Government* 1975. 189 pp.

170. EARLE, CARVILLE V. *The Evolution of a Tidewater Settlement System: All Hallow's Parish, Maryland, 1650–1783* 1975. 249 pp.

171. SPODEK, HOWARD. *Urban-Rural Integration in Regional Development: A Case Study of Saurashtra, India—1800–1960* 1976. 156 pp.

172. COHEN, YEHOSHUA S. and BERRY, BRIAN J. L. *Spatial Components of Manufacturing Change* 1975. 272 pp.

173. HAYES, CHARLES R. *The Dispersed City: The Case of Piedmont, North Carolina* 1976. 169 pp.

174. CARGO, DOUGLAS B. *Solid Wastes: Factors Influencing Generation Rates* 1977. 112 pp.

175. GILLARD, QUENTIN. *Incomes and Accessibility. Metropolitan Labor Force Participation, Commuting, and Income Differentials in the United States, 1960–1970* 1977. 140 pp.

176. MORGAN, DAVID J. *Patterns of Population Distribution: A Residential Preference Model and Its Dynamic* 1978. 216 pp.

177. STOKES, HOUSTON H.; JONES, DONALD W. and NEUBURGER, HUGH M. *Unemployment and Adjustment in the Labor Market: A Comparison between the Regional and National Responses* 1975. 135 pp.

179. HARRIS, CHAUNCY D. *Bibliography of Geography. Part I. Introduction to General Aids* 1976. 288 pp.

180. CARR, CLAUDIA J. *Pastoralism in Crisis. The Dasanetch and their Ethiopian Lands.* 1977. 339 pp.

181. GOODWIN, GARY C. *Cherokees in Transition: A Study of Changing Culture and Environment Prior to 1775.* 1977. 221 pp.

182. KNIGHT, DAVID B. *A Capital for Canada: Conflict and Compromise in the Nineteenth Century.* 1977. 359 pp.

183. HAIGH, MARTIN J. *The Evolution of Slopes on Artificial Landforms: Blaenavon, Gwent.* 1978. 311 pp.

184. FINK, L. DEE. *Listening to the Learner. An Exploratory Study of Personal Meaning in College Geography Courses.* 1977. 200 pp.

185. HELGREN, DAVID M. *Rivers of Diamonds: An Alluvial History of the Lower Vaal Basin.* 1979. 399 pp.

186. BUTZER, KARL W., *editor. Dimensions of Human Geography: Essays on Some Familiar and Neglected Themes.* 1978. 201 pp.

187. MITSUHASHI, SETSUKO. *Japanese Commodity Flows.* 1978. 185 pp.

188. CARIS, SUSAN L. *Community Attitudes toward Pollution.* 1978. 226 pp.

189. REES, PHILIP M. *Residential Patterns in American Cities, 1960.* 1979. 424 pp.

190. KANNE, EDWARD A. *Fresh Food for Nicosia.* 1979. 116 pp.